THE CONSERVATION OF AQUATIC SYSTEMS

Proceedings of a seminar held on 18–19 February 1993

edited by

Julian D. Reynolds

Department of Zoology, Trinity College, Dublin

Royal Irish Academy
Dublin 1996

ROYAL IRISH ACADEMY
19 Dawson Street
Dublin 2

Cover design: PHILIP QUESTED

Cover photograph of Lough Bray, Co. Wicklow,
a dystrophic cirque lake affected by acid precipitation: JULIAN D. REYNOLDS

ISBN 1 874045 22 4

British Library Cataloguing-in-Publication Data.
A catalogue record for this book is available from the British Library

Also published by the Royal Irish Academy

Renewable resource development in Ireland
edited by A.C. Cassells

Biological indicators of pollution
edited by D.H.S. Richardson

Taxonomy: putting plants and animals in their place
edited by Christopher Moriarty

Ecology and conservation of Irish peatlands
edited by Gerry Doyle

Irish rivers: biology and management
edited by Martin Steer

Plant health and 1992
edited by J.A. Kavanagh and P. Brennan

The badger
edited by Thomas J. Hayden

Irish grasslands: their biology and management
edited by D.W. Jeffrey, M.B. Jones and J.H. McAdam

Wood, trees and forests in Ireland
edited by Jon R. Pilcher and Seán Mac an tSaoir

CONTENTS

CONTRIBUTORS

ANDREW, T.E.	Department of Biological and Biomedical Sciences, University of Ulster, Coleraine BT52 1SA, Northern Ireland
BOADEN, P.J.S.	School of Biology and Biochemistry, Queen's University of Belfast Marine Biology Station, Portaferry, Co. Down BT22 1PF
BOWMAN, J.	Environmental Protection Agency, Pottery Road, Dun Laoghaire, Co. Dublin
BRACKEN, J.J.	Department of Zoology, University College, Belfield, Dublin 4
CHESNEY, HELENA G.C.	Ulster Museum, Botanic Gardens, Belfast BT9 5AB
CLABBY, K.C.	Environmental Protection Agency, Regional Inspectorate, Pottery Road, Dun Laoghaire, Co. Dublin
DONNELLY, R.E.	Department of Zoology, University of Dublin, Trinity College, Dublin 2
GRAINGER, J.N.R.	Department of Zoology, University of Dublin, Trinity College, Dublin 2
GRIFFITH, DAVID de G.	Fisheries Research Centre, Abbotstown, Dublin 15.
HICKIE, DAVID	An Taisce—The National Trust for Ireland, Tailors Hall, Dublin 8
KEEGAN, BRENDAN F.	Department of Zoology, The Martin Ryan Marine Science Institute, University College, Galway
KELLY-QUINN, M.	Department of Zoology, University College, Belfield, Dublin 4
KING, PAULINE	Department of Zoology, The Martin Ryan Marine Science Institute, University College, Galway
LAWLER, IAN	Department of Zoology, University of Dublin, Trinity College, Dublin 2
LUCEY, J.	Environmental Protection Agency, Regional Inspectorate, Kilkenny
LUNNON, RUTH	Department of Zoology, University of Dublin, Trinity College, Dublin 2

MAITLAND, PETER S.	Conservation Centre, Easter Cringate, Stirling FK7 9QX, Scotland
MARNELL, FERDIA	Department of Zoology, University of Dublin, Trinity College, Dublin 2
MATTHEWS, MILTON A.	The Salmon Research Agency of Ireland, Farran Laboratory, Furnace, Newport, Co. Mayo
McGARRIGLE, M.L.	Environmental Protection Agency, Regional Inspectorate, Castlebar, Co. Mayo
KcKELVEY, F.B.	Department of Biological and Biomedical Sciences, University of Ulster, Coleraine BT52 1SA, Northern Ireland
MURRAY, DECLAN A.	Department of Zoology, University College, Belfield, Dublin 4
O'SULLIVAN, W.M.	Department of Zoology, University College, Cork
POOLE, W.R.	The Salmon Research Agency of Ireland, Farran Laboratory, Furnace, Newport, Co. Mayo
REYNOLDS, JULIAN D.	Department of Zoology, University of Dublin, Trinity College, Dublin 2
RYAN, J.M.	National Parks and Wildlife Service, Office of Public Works, Dublin 2
TIERNEY, D.	Department of Zoology, University College, Belfield, Dublin 4
TULLY, OLIVER	Department of Zoology, University of Dublin, Trinity College, Dublin 2
WALSH, T.	Department of Zoology, University of Dublin, Trinity College, Dublin 2
WHELAN, CAROLINE	Department of Zoology, University of Dublin, Trinity College, Dublin 2
WHELAN, K.F.	The Salmon Research Agency of Ireland, Farran Laboratory, Furnace, Newport, Co. Mayo
WILSON, JAMES G.	Environmental Sciences Unit, University of Dublin, Trinity College, Dublin 2

In: J.D. Reynolds (ed.) 1996 *The conservation of aquatic systems*, 1–2. Dublin. Royal Irish Academy.

INTRODUCTION

The genesis of this seminar came in 1990 through the Royal Irish Academy's Sub-committee for Hydrobiology, when I was asked to produce suggestions for a meeting on conservation of Irish aquatic resources. This allowed me to develop ideas that had been gestating for several years, and in 1991 a small committee, consisting of Dr D. Murray (Department of Zoology, University College, Dublin), Dr C. Moriarty (Fisheries Research Centre, Department of the Marine), Dr J. Wilson (Environmental Sciences Unit, Trinity College, Dublin) and myself, was formed to plan and develop the detailed programme, to select speakers and chairmen and to bring the seminar to reality—and for all their help I am very grateful.

The seminar was opened by Mr Gerry O'Sullivan, TD, minister of state for the Department of the Marine, who reviewed the development of marine research and conservation and acknowledged the pioneering role of the Royal Irish Academy in aquatic research: its exploration of the deep waters over the continental shelf may be seen as the first major development in the aquatic sciences in Ireland. This was followed quickly by the Royal Dublin Society's work on the exploitable fishes of our inshore waters under W.S. Green, which led directly to the establishment of Irish fisheries science. The emphasis then was on exploitation rather than conservation, but the recognition of the need for careful biological groundwork allowed for stock management, which is conservation. Irish workers were among the first to be involved in the International Council for the Exploration of the Sea, one of the oldest international scientific bodies in the world. In freshwaters, early detailed studies also laid down valuable baselines.

Our geographical position exposes us to international stresses and potentially serious impacts of marine accidents and oil spills, and so the world impinges on Irish aquatic systems. Against this background the minister reported that Ireland this year called for urgent steps to be taken by the European Commission to minimise the risk of such occurrences. Tourism and fisheries are both dependent on the health of the environment. We have been spared the worst abuses resulting from industrialisation in other countries, and the clean environment we enjoy has both a cash value and an aesthetic one.

What we do with aquatic systems in Ireland also has wider repercussions. This seminar helps to focus our attention on the national and international nature of the problem, and the minister emphasised that the state of the environment can be maintained only by eternal vigilance. He wished success to the deliberations of the seminar and looked forward to the fruits of its work leading to better conservation of our aquatic systems.

The seminar consisted of twelve invited papers, and a number of offered posters presented in an interactive session. All major papers have been subjected to peer review and are arranged in this book under four sections: reviews and case studies of Irish marine and freshwater systems; studies of the conservation value of selected aquatic organisms; the major impacts that may constrain conservation; and approaches to conservation as evaluated by government and non-governmental agencies.

I am most grateful to Ms Sara Whelan, Dr Rachel Finnegan and the secretarial staff of the Royal Irish Academy for their smooth administration of the meeting, and to all the chairmen, speakers and referees for their diligent work and for helping to make this seminar a valuable reflection of the current state of conservation in Irish aquatic systems. Generous sponsorship by the Department of the Environment, by the Department of the Marine and by the Electricity Supply Board has made publication of this seminar possible.

Julian D. Reynolds
Seminar Coordinator and Editor

In: J.D. Reynolds (ed.) 1996 *The conservation of aquatic systems*, 3–19. Dublin. Royal Irish Academy.

FRESHWATER RESOURCES, SYSTEMS AND HABITATS IN IRELAND

Declan A. Murray

Department of Zoology, University College, Dublin

ABSTRACT

Water is essential to sustain life but substantially less than 1% of the world's water resource is freshwater. It is this fraction that sustains all terrestrial life and also provides habitats for myriads of freshwater organisms. Ireland's renewable water resources and freshwater habitats are maintained by a mean annual precipitation of approximately 1150mm, about 38% of which is consumed by evaporation and transpiration. The remaining 62% of rainfall (about 59km^3) constitutes the annual average surface run-off supplying the streams, rivers, lakes, wetlands and groundwater aquifers in Ireland. It is from these systems that water is abstracted for human use, the first four providing the main freshwater habitats for resident aquatic organisms and migratory birds. This paper presents a brief overview of these aquatic habitats and addresses some current problems in their conservation.

INTRODUCTION

Water is such a ubiquitous substance that, like the air we breathe, we take it for granted. The phrases 'water of life', 'water is life' and 'water for life' have been repeated so often and in so many different languages—'uisce beatha', 'aqua vitae' and 'akvavit'—that unfortunately they have become catch-phrases rather than giving proper recognition of their real meaning. Water means different things to different people at different times. It means drink, food and survival when it comes at the right time, in the right amount and in the right place. It may bring floods and disaster when too plentiful, or drought and famine when in short supply. It is a central component of the human social and economic system: it is used in agriculture, in the provision of electricity for our homes and industries, for transport, and for recreational pleasure in boating, fishing and swimming. It is a powerful solvent enabling living organisms to use aqueous solutions, such as sap, blood and digestive juices, to maintain their biological processes. Water is a vital substance; without it life as we know it could not exist.

Water is a finite resource existing in different phases of the hydrological cycle driven by the action of solar energy and gravity in an exchange between the oceans, the atmosphere and land. A recent estimate puts the total volume of water in its different phases on our planet at 1386 million km^3 (Shiklomanova 1990). Approximately 96.5% of this water is saline, and the total volume of freshwater in the hydrosphere is about 35 million km^3, 69% of which occurs in the form of ice and snow in polar regions. Groundwater accounts for a further 30% of the freshwater resource, while freshwater lakes and rivers contain only 93,000 km^3, which is approximately 0.0072% of the total amount of water on earth. This relatively small proportion of freshwater plays a disproportionately large role in our lives.

THE IRISH FRESHWATER RESOURCE

Ireland's freshwater comes predominantly from the prevailing south-west/west winds, blowing across and collecting moisture from the Atlantic for most of the year and delivering an annual average precipitation of 1150mm over the land mass of 84,420km^2 (McCumiskey 1991). About 38% (450mm) of this precipitation is 'lost' to the atmosphere by evapotranspiration, leaving a net annual surface accumulation of 700mm or an overall volume of approximately 59km^3. This annual run-off sustains the freshwater habitats for many different resident and migratory organisms and supplies the requirements for human consumption. Estimated at 11,370m^3 per capita per annum (McCumiskey 1991), this freshwater resource is 4–5 times that of the more densely populated and industrialised neighbouring countries in the European Community, where per capita water availability ranges from 1260m^3/yr per person in Belgium to 6380m^3/yr per person in Luxembourg (ECOTEC 1992).

IRELAND'S FRESHWATER SYSTEMS AND HABITATS

The network of rivers, lakes, bogs, wetlands and groundwater aquifers in Ireland has been dictated by geological history. Ireland's geological map, a mixture of sedimentary and metamorphic rocks, granite, sandstone and carboniferous deposits, shales and basalt, is a patchwork arising from a 500-million-year history (Kennan 1992). Two periods of glaciation during the Pleistocene are evident in the country's glacial deposits, ice-rounded rocks, valleys and lakes. Since the final retreat of ice, about 10,000 years ago, water and natural weathering processes have continued, through erosion and leaching, to shape the landscape. The broad central lowland, interrupted by areas of plateau, is about twice as long from east to west as it is from north to south and is dominated by glacial drift materials of clay, sand and gravel with considerable areas of wetland, raised bog and lakes. Two ridges of higher ground run from north to south, one extending from Monaghan to the Slieve Bloom mountains and the second from east of the Ox mountains to Slieve Aughty in south-east Galway. The plain is surrounded by discontinuous mountain ranges, seldom reaching more than 600m. In the north-east of the country there is a plateau of basalt, while the mountains of the south and south-west are sandstone with limestone deposits in their river valleys. Granite dominates in the north-west, in the mountains of Donegal, Mayo and Galway, and in the east, in Dublin and Wicklow. It is against this background that the component parts of the Irish freshwater resource may be assessed.

Flowing (lotic) waters

Precipitation runs off or percolates through soil to emerge as a spring—the genesis of a dynamic, unidirectional lotic system relentlessly progressing under gravitational forces to the sea. Rivers arising on the seaward side of Ireland's mountain rim drain small catchments and flow rapidly and directly to the sea. Those arising on the landward side, or in the central lowlands, flow to the sea at a more leisurely pace, often through lakes and through a variety of glacially deposited soil types overlying different geological strata. Of the 75 or so river catchments over 130km^2 in area, the Shannon, draining an area of 10,400km^2 and lying between the two north–south ridges of the central lowland, is the largest (Fig. 1). The northern part of the lowlands is drained by the Erne, flowing slowly through numerous lakes and drumlin terrain to exit to the sea in Donegal Bay. The principal catchments of the central lowland east of the ridge extending from Monaghan to the Slieve Bloom mountains are drained by the rivers Boyne, Liffey, Slaney, Barrow, Nore and Suir, while land west of the ridge, between Sligo and south Galway, is drained by the rivers Moy and Corrib. The rivers Blackwater, Lee, Bandon and Laune drain the major catchments of the south and south-west, with the rivers Newry, Lagan, Bush, Bann and Foyle draining lands in the north-east and north.

Erosion is the dominant feature in lotic waters, in which longitudinal changes in ecology from the source of a river to where it enters the sea may be extreme, especially over large elevation gradients. In addition to the visible differences in gradient, physical characteristics and substrate type along the length of a watercourse, there are also differences on a smaller scale within a river that are related to variation in current velocity, oxygen content and build-up of organic materials. Furthermore, the chemical composition of water, determined primarily by bedrock and soils in the catchment, is a major factor influencing the type and distribution of organisms within lotic systems. Thus, waters lying in or flowing through Ireland's different geological strata of granite, sandstone or calcareous deposits each have a characteristic flora and fauna. Rivers and streams in the mountains of Antrim, Donegal, Mayo, Galway, Kerry, Cork and Wicklow are nutrient deficient, but their productivity varies with the underlying bedrock. The granites of Donegal and Wicklow are more nutrient poor than the basalts of Antrim or the slates of south Mayo, all of which are inherently low in nutrients such as phosphates and nitrates.

Many streams in mountain regions originate in blanket bogs, which give

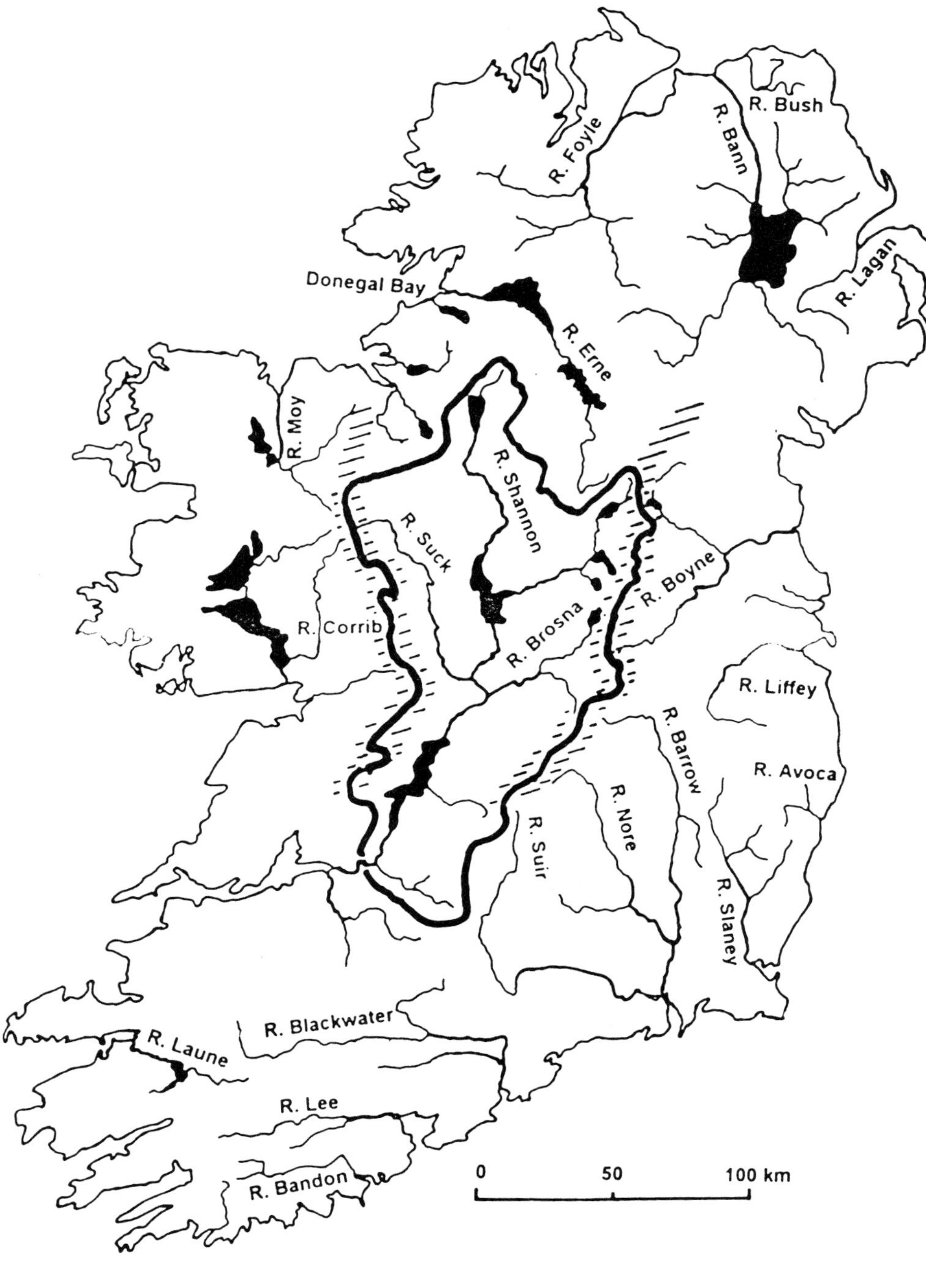

Fig. 1—Major catchments in Ireland. Solid lines outline the River Shannon catchment, hatched lines indicate the two north–south ridges (see text).

a distinctive brown colour to these slightly acidic and oligotrophic headwaters. Rivers in upland reaches may present an immense variety of habitats. Although current velocities are generally unfavourable for the development of a pleuston (surface dwelling) community, such a community, represented by the Gerridae (water strider), various species of whirligig beetles (Gyrinidae) and the meniscus midge (Dixidae), living in foam aggregations, may occur in well vegetated stream margins and backwaters. Madicolous organisms, dominated by insects like the Chironomidae, Ceratopogonidae, Psychodidae and Thaumalidae, are found on the thin film of water flowing over rock faces, on splash zones and on seep trickles. Substrates of gravels, partially and fully submerged rocks and boulders, with waterfalls, plunge pools and rapids interspersed by more tranquil, slow-flowing stretches, each provide opportunities for colonisation by a different benthos. A highly specialised benthic community, with morphological and physiological adaptations to the prevailing conditions of lower temperatures, fast current and abundance of oxygen, lives in these upper reaches. Insects, including the Trichoptera, Plecoptera, Ephemeroptera, Chironomidae and Simuliidae, are dominant except in upland limestone regions, where crustaceans and other invertebrates may be more plentiful. Headwater streams provide the winter spawning grounds for fish such as salmon and trout and the nursery grounds for their young, each competing for food and space.

In contrast to the rivers in upland granite and shale regions of the country, waters traversing the glacial deposits of the lowlands tend to be sluggish, nutrient rich and more productive. However, even where the gradient is gentle, a sequence of erosion and deposition occurs, giving riffles, glides and pools and opportunities for aquatic plants and animals to exploit these habitats. Temperature fluctuations are greater here in comparison with headwater regions and oxygen concentrations are more variable and may be lower. Different substrates of boulders, stones, sand, silt and mud determine the composition of the benthic communities. The surfaces of boulders and stones are colonised by mosses which in turn provide habitats for many insect species. Loose sandy substrates are poor habitats for invertebrates because of their instability and their inability to hold organic matter. *Cryptochironomus* larvae of the *Harnischia* complex (Chironomidae) prefer sandy substrates and are commonly found in such habitats in Irish rivers. Rivers with gravel beds also possess a deeper interstitial habitat formed in the mixture of coarse sand, gravel and rocks. Little work has been carried out on such hyporheic habitats in Ireland, which contain elements of the surface benthos, such as chironomid and beetle larvae, mayflies and stoneflies. The hyporheic zone has a lateral component and may extend under the stream bank. Ward (1989) recorded stonefly nymphs up to 50m from the main channel in loosely-compacted gravels in Montana, USA.

Rivers flowing through the calcareous waters of the midlands are naturally productive and support a variety of invertebrates such as Trichoptera, crustaceans (*Gammarus*, *Asellus* and the decapod crayfish *Austropotamobius*), molluscs, coleopterans, Diptera, etc. Removal of CO_2 from water by algae in these limestone-rich regions results in deposition of calcium in characteristic formations of 'tufa', which provide shelter for a range of organisms including caddisfly larvae, chironomids and beetles.

Lentic waters—lakes

There are over 4000 lakes in Ireland, mostly of glacial origin, which with a combined surface area of approximately 168,000ha account for about 2% of the country's total surface area. The majority of these are small and situated in less densely populated regions, along the western seaboard from Kerry to Donegal. A number of large lakes in the Shannon catchment and a variety of smaller water bodies in the drumlin region of Cavan and Monaghan drain through the River Erne. Some 23 lakes have surface areas in excess of 1000ha, minute by comparison with Lake Superior, the largest lake in the world, whose surface area is only slightly less than the entire area of Ireland (Fig. 2, Table 1). An additional 45 or so range from 100ha to 1000ha, while the remainder are less than 100ha. Most Irish lakes are shallow and depths in excess of 50m are rarely encountered. The country's largest lake, Lough Neagh, has a surface area of approximately 38,300ha and a maximum depth of 33m, but has a mean depth of only 8m.

Collectively, lakes offer a diverse array of habitats in different depth zones and on varying substrata. Lake zonation schemes, primarily based on rooted macrophytes, demarcate habitats in the littoral, sublittoral and profundal zones, while in the open water area pleustonic, planktonic and nektonic communities are found. Ecological conditions related to substrate type and degree of exposure in the littoral zone vary from lake to lake and at different locations within the same lake. Thus, the exposed rocky shore

Table 1—Lakes in Ireland with surface area greater than 730ha.

Lake	*Area ha*	*Catchment*	*Lake*	*Area ha*	*Catchment*
Neagh	38,800	Bann	Gill	1,400	Garavogue
Corrib	17,000	Corrib	Gowna	1,290	Erne
Derg	11,600	Shannon	Arrow	1,280	Unshin
Erne, lower	10,950	Erne	Ennell	1,200	Brosna
Ree	10,500	Shannon	Currane	1,100	Cumeragh
Mask	8,300	Corrib	Gara	1,100	Boyle
Conn	5,000	Moy	Oughter	1,035	Erne
Allen	3,500	Shannon	Cullin	1,000	Moy
Erne, upper	3,450	Erne	Derravaragh	1,000	Inny
Melvin	2,100	Drowes	Owel	950	Brosna
Leane	2,000	Laune	Key	900	Boyle
Blessington (R)	1,970	Liffey	Carrigadrohid (R)	900	Blackwater
Sheelin	1,900	Inny	Ramor	737	Boyne
Carra	1,500	Corrib			

(R= reservoir)

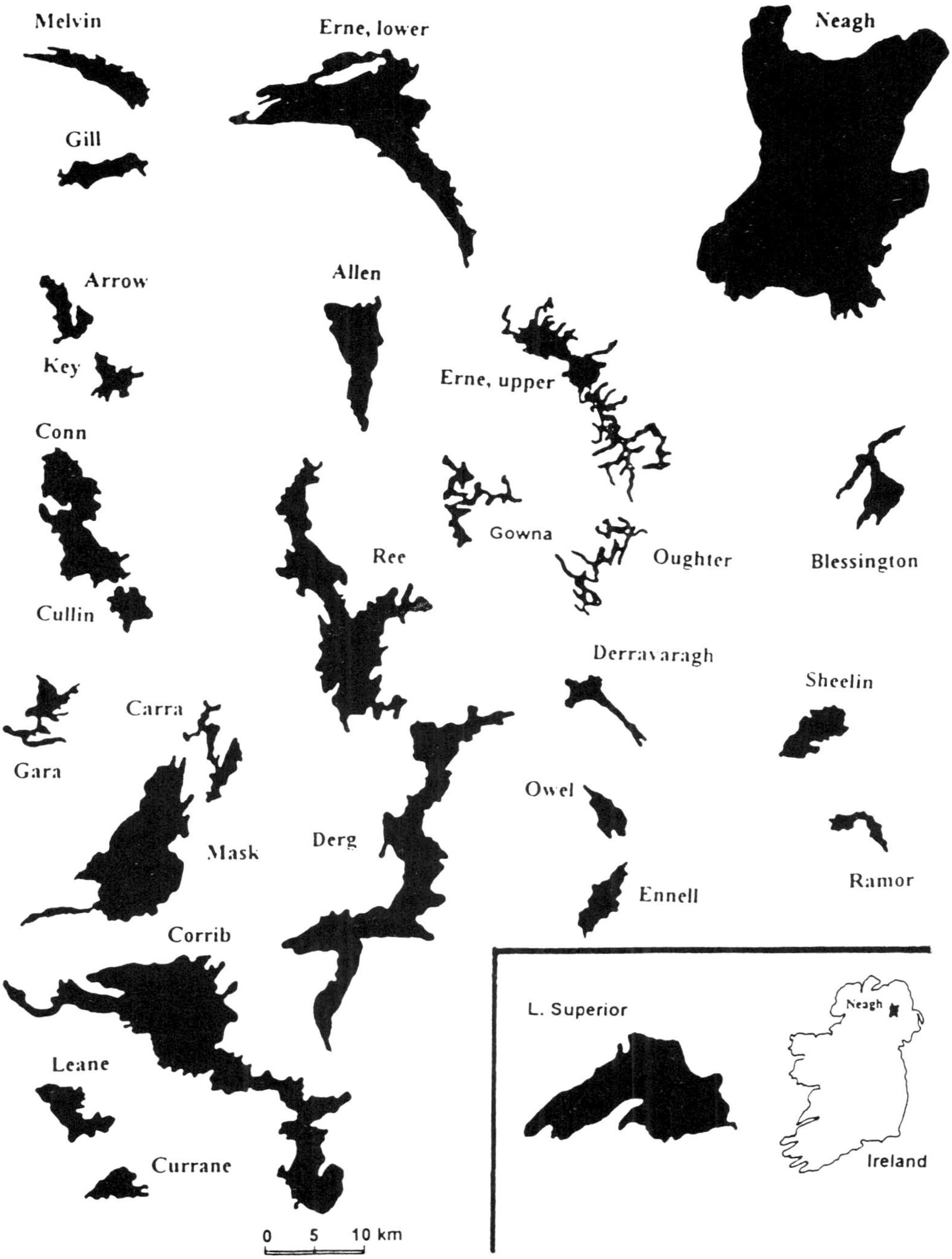

Fig. 2—Comparison of surface areas of Ireland's largest lakes. Inset depicts Lake Superior, USA, and Ireland drawn to the same scale.

with coarse substrate, continual water movement and silt-free interstices differs from the sheltered shore whose stands of emergent vegetation provide a high habitat diversity with a rich fauna. The sublittoral zone contains fauna from the lower limits of the littoral and upward extensions of the profundal zones. The profundal zone, with a substrate of fine soft ooze, is dominated by various species of chironomid larvae and oligochaetes. The abundance of certain species of chironomid larvae in the profundal sediments is often characteristic of lake type, and classification schemes based on this fauna have application in Irish lakes.

Since a lake is an integral part of the watershed in which it occurs, variation in the water quality and productivity of Irish lakes results from a combination of natural and anthropogenic influences. Mountain lakes in Donegal and Wicklow lying in nutrient-poor bedrock tend to be oligotrophic and to support a characteristic flora and fauna. Lakes in the limestone areas of the midlands and west, such as loughs Sheelin, Ennell, and Derravaragh, and loughs Carra, Corrib, Mask and Conn, are naturally productive, well endowed with diverse biotic communities and important as trout angling resources. Because Irish lakes are generally shallow and wind-swept, stratification is not necessarily a regular annual event. They may be categorised as warm monomictic, when stratification develops once during the summer months, or polymictic if the lake water stratifies on more than one occasion in the summer, or pleomictic if weather conditions are not conducive to the development of stratification. Oxygen depletion or complete anoxia of the hypolimnion is a common feature during summer stratification in continental eutrophic lakes. Hypolimnetic anoxia does not usually occur in most productive Irish lakes because stratification (if it occurs) is rarely so prolonged as to bring about complete anoxia under the thermocline. However, oxygen depletion occurs and has been recorded in some western lakes while hypolimnetic anoxia was observed in Lough Ennell during the decay of a dense algal bloom (Champ 1977).

Wetlands

Wetlands, 'areas of marsh, fen, peatland or water, whether natural or artificial, permanent or temporary, with water that is static or flowing, fresh, brackish or salt including areas of marine water the depth of which at low tide does not exceed six metres' (Anon. 1971), perform different functions in relation to water management. They influence recharge and discharge of groundwater, control floods and remove nutrients and pollutants. Irish freshwater wetlands are fens, raised bogs, blanket bogs (Atlantic and mountain), river flood-plains or callows and turloughs. All provide opportunities for a variety of aquatic and semi-terrestrial flora and fauna and each, with its own particular ecology, provides habitats for specialised plant and animal species adapted to withstand the extreme conditions (Doyle 1990; 1992).

Fens, alkaline peat lands that occur in waterlogged low-lying midland regions overlying limestone, are fed by calcareous-rich waters. Consequently fens have a rich and varied vegetation with a greater diversity than bogs and with open water and marginal calcium-rich aquatic habitats for molluscs, chironomids, Odonata and hemipterans. Atlantic and mountain blanket bogs and raised bogs derive their nutrient supply from rainwater and have acidic surface peat. Frequent high winds and low temperatures make blanket bogs inhospitable regions; nevertheless a variety of organisms such as odonatan nymphs and megalopteran or chironomid larvae and Cladocera live in such bogs in small lakes, ponds, seepages and wet tussocks (Dowling and Murray 1981; Duigan 1992). Most notorious in these water bodies are the ceratopogonid larvae, whose adult stage, the biting midge, can make field work in bogs almost unbearable in summer months. Raised bogs are more typically found in midland regions and have pools, seepages and soaks inhabited by a variety of invertebrates (Reynolds 1986a). Some highly specialised species such as the chironomid *Lasiodiamesa sphagnicola* live in the wet foliage of *Sphagnum*, a characteristic bog moss (Ashe 1987).

An extensive area of river floodplain adjoins the Shannon between Athlone and Portumna and includes the lower parts of the rivers Suck and Little Brosna to form the wet grasslands known as 'callows', a region of treeless fen peat bordering and periodically inundated by flood water. Temporary 'lakes' are formed in the callows during high water conditions of winter and are important refugia for overwintering wildfowl such as whooper swan, white-fronted goose, widgeon, golden plover and godwit.

West of the Shannon, in the limestone karst region of south Galway and Clare, variation in groundwater levels is reflected in episodic surface water accumulations. Here Ireland's seasonal winter lakes, or turloughs, occur on glacial drift material. The turlough at Rahasane, Co. Galway, is the largest in the country, with a winter-time surface area of 250ha. Such seasonal waters have a characteristic crustacean fauna, including *Tanymastix* and a glacial relict species, *Eurycercus glacialis,* adapted to survive periods of drought (Duigan 1992; Duigan and Frey 1987; Reynolds, 1986b).

Ground water

Surface and ground waters are inextricably linked. Surface water percolates through the soils and underlying strata under the force of gravity until it discharges into a spring, stream, wetland, lake or to the sea. Ground water (*stygon*) comprises an enormous subterranean body of water providing habitats between the grains of fine sand *(psammal*), coarse sand (*psephal*), in caves (*troglon*) or emerging to the surface in springs (*crenon*). The stygon environment is very constant in contrast to surface-water ecosystems. There is little seasonal variation in temperature or oxygen concentrations and light is permanently absent. For living organisms this is a hostile environment where, according to Illies (1974), old (Tertiary relict) faunal elements live. The eyeless amphipod *Niphargus kochianus* has been recorded from caves in the Burren, Co. Clare, and from wells. Since light is absent in these habitats there is no primary productivity and these animals feed on detritus materials transported in suspension from the surface waters.

MANAGING THE RESOURCE

Almost all human activities have an impact on freshwater resources and in most instances lead to their degradation. Freshwater is occasionally overused and frequently wasted. Ireland is fortunate to have an abundance of rivers, lakes and wetlands, the majority of which are as yet relatively unpolluted. However, there are conservation issues related to each of the freshwater habitats addressed above.

Conservation issues for lotic habitats

Erosion is common in upland regions of mountain blanket bog resulting from a combination of factors associated with climate but compounded by human activities of turf cutting, afforestation and over-grazing by hill sheep (Bradshaw 1987). Such erosion gives rise to peat-silt inwash to streams, which has been implicated in damage to some fisheries in western counties (Salmon Research Agency 1990).

Artificial acidification is a serious problem in many parts of Europe, particularly Scandinavia. Some systems are naturally acidic owing to inwash of peat-silt with high levels of humic acids, a phenomenon exacerbated by erosion. Afforestation with evergreen trees has also been implicated in the observed decrease in pH of some waters in mountain streams in poorly buffered regions of the country (Bowman 1991; Reynolds *et al.* 1994; Kelly-Quinn *et al.,* this volume).

Arterial drainage has been carried out in a number of river systems in Ireland. Studies undertaken on the effects of such activities in river channels have shown that the ecosystem is initially drastically changed especially in the composition and distribution of flora, macroinvertebrate fauna and fish (Toner *et al.* 1965; McCarthy 1977). Recovery of the fauna and flora to pre-drainage conditions is dependent on, among other factors, gradient, substrate type, bank material and the capacity for re-establishment of a riparian treeline and shrubbery. Instances of recovery have been documented by Kennedy (1980) on the River Camowen, by McCarthy (1983) and O'Grady (1991) on the Trimblestown river, and by O'Grady and King (1992) on the Bunree river. However, in low-gradient systems recovery may not occur naturally and post-drainage, in-stream enhancement procedures have been successfully undertaken to improve invertebrate and fish populations (O'Grady 1990; Lynch and Murray 1994).

The most widespread cause of deterioration of river water quality in Ireland is organic enrichment—eutrophication—from diffuse sources (agriculture) and point sources, mostly sewage effluent but also 'accidental' spills of toxic materials, including silage effluent. These inputs may lead to degradation of the habitat, oxygen depletion and fish kills. McCumiskey (1991) notes that the effluent from a number of Irish towns (39 out of 128 towns surveyed) receives no treatment prior to discharge. The bulk of the untreated discharges are made to the sea but some are directly to rivers. The EC Directive on urban wastewater treatment (91/271/EC) will undoubtedly bring about a reduction in such point source discharges but the problem of diffuse discharges will remain. The Directive also imposes restrictions on the disposal of sludge to the sea. Casey and Dowley (1992) review the implications of this directive.

The most recent study on 13,470km of river by the Environmental Research Unit over the period 1987–90 indicated that approximately 76.4% of the channel length examined was unpolluted, 12.1% slightly polluted, 10.3% moderately polluted and 1.2% (162km) of river channel was seriously polluted (McGarrigle *et al.* 1992 and this volume). Comparable results in Northern Ireland on a channel length of 1744 km indicated that 57% was unpolluted, 35% slightly polluted, 8% moderately polluted and 1% seriously polluted (Sweeting *et al.* 1992). The extent of pollution is related to the degree of human impact and is thus virtually non-existent in the more remote districts of the country. More than 94% of the surveyed channel in the Donegal–Sligo region was unpolluted, while in the more densely populated eastern region the pollution-free proportion was 60%. Only 46.5% of the river and stream channels surveyed in the Cavan–Monaghan region were unpolluted. Comparison with results from earlier surveys (Table 2) shows that since 1971 the incidence of serious pollution diminished significantly but there was an overall increase in the extent of pollution on the 2900km of river channel included in the first baseline survey. This was attributed to eutrophication causing an increase in channel length exhibiting signs of slight and moderate pollution. A similar trend is evident in the longer baseline for the periods 1982–6 and 1987–90.

Conservation issues for lakes

Investigations by Bowman (1991) on lakes in sensitive, i.e. poorly buffered, areas in Ireland have shown that the incidence of acidification from precipitation is minimal. Irish lakes have been spared the toxic pollution associated with industrialisation but have not escaped the problems of cultural eutrophication, which became a cause for concern in the 1970s. Lough Neagh was ranked among the most eutrophic lakes in Europe (Wood and Gibson

Table 2—Trends in river water quality for the period 1971–90 expressed as percentage of channel length in four water quality categories for a 2900km channel length in the years 1971, 1981, 1986 and 1990 and for a 6900km channel length for the periods 1982–86 and 1987–90 (from McGarrigle *et al.* 1992).

	1971	*1981*	*1986*	*1990*	*1982–86*	*1987–90*
Quality status %						
Unpolluted	83	78	69	65	74	70
Slightly polluted	5	11	20	20	15	16
Moderately polluted	5	7	8	13	9	12
Seriously polluted	7	4	3	2	2	1.5

1973). At the same time a number of trout fishery lakes, including loughs Ennell and Sheelin (Champ 1977) and Lough Leane (Bracken *et al.* 1977), were identified as exhibiting signs of accelerated eutrophication caused by artificial enrichment either through point-source sewage discharge or diffuse input from agriculture. With the realisation that cultural eutrophication may be halted, alleviated and in some cases reversed, by controlling phosphorous inputs from point sources, the water quality in lakes such as Lough Ennell and Lough Leane has improved in recent years (Clabby *et al.* 1992). Similarly some recovery in Lough Neagh has to date ensured the survival of a valuable eel fishery (Wood 1989). However, artificial enrichment remains a matter for concern today.

Eighty percent of the 172 lakes examined in the recent survey by the Environmental Research Unit were categorised as being either oligotrophic or mesotrophic, consistent with a low probability of pollution (Clabby *et al.* 1992). Fourteen lakes were categorised as moderately eutrophic, fifteen were either strongly or highly eutrophic (including Lough Derg and Carrigadrohid Reservoir) and six, including Lough Ramor, Co. Cavan, were hypereutrophic. Most of the lakes exhibiting high levels of algal growth are relatively small and lie in the north-midland counties of Cavan, Leitrim and Monaghan, where enrichment is primarily associated with non-point discharge of nutrients derived from agricultural activities. Control of phosphorous inputs from diffuse agricultural sources is difficult. Sherwood (1992) highlighted the need for good management practices to protect aquatic resources from runoff of slurries following land spreading by avoiding application when rainfall is imminent. Management of slurries in the Lough Sheelin catchment has brought about a marked improvement in water quality and a recovery of trout stocks (Duggan and Champ 1992).

Water quality in Lough Derg, the lowermost lake on the Shannon and the final sedimentation trap of the Shannon catchment, has progressively deteriorated in recent years. Ongoing palaeolimnological (personal) studies indicate that sediments are currently accumulating at a rate of approximately 6mm per year in the lower basin. The results of investigations by the Environmental Research Unit on nutrient inputs to the lake are presented by Bowman (this volume).

Conservation issues for wetlands

The principal threats to aquatic habitats in peatlands are exploitation for turf cutting, afforestation, drainage of marginal regions for land reclamation and overgrazing by sheep of mountain blanket bog. Doyle (1992) notes that peatland exploitation continues at a rapid pace and concludes that a well planned acquisition programme is needed to save an important set of Irish peatlands. Arterial drainage is the prime threat to turloughs and other wetlands. Nineteenth-century drainage reduced the extent of turloughs in the country by about one third and today local drainage threatens to alter some remaining turloughs, including Rahasane turlough, part of the Dunkellin River catchment in County Clare.

Conservation issues for groundwater

Groundwater is a major renewable resource, providing 20%–25% of the drinking water supplies in Ireland (Daly 1992). However, since groundwater abstractions at present account for only 3% of the estimated annual recharge, the perception is that Ireland's vast groundwater resources are currently under-used. Increasing quantitative demands on this resource could have a detrimental impact on surface-water resources through lowering of the groundwater table, reducing the stream summer base flow and reducing water inflows to fens and turloughs. Some control on this potential threat is vested in an EC Directive (1985), which requires that an Environmental Impact Statement be prepared for any proposed abstraction exceeding 5000m^3 per day. Ireland's groundwaters are also threatened by anthropogenic contaminants, principally faecal bacteria, ammonia, nitrate, potassium, chloride, iron and manganese (Daly 1992). Point sources, such as organic wastes in farmyards, septic tank systems and landfill sites, have been implicated in some instances, but it is also clear that sources of groundwater contamination may be more widespread and related to land spreading of organic wastes, organic fertilisers, pesticides and urban run-off (Daly and Daly 1984; Thorn 1986; Neill 1989). Thorn and Coxon (1992) have reviewed the situation in relation to the 1991 EC Nitrate Directive.

THE WAY FORWARD

In January 1992 the United Nations International Conference on Water and the Environment (ICWE) was held in Dublin in preparation for Earth Summit in Brazil (Anon. 1992). Attended by delegates from over a hundred countries, the meeting culminated with the Dublin Statement:

> Scarcity and misuse of freshwater pose a serious and growing threat to sustainable development and protection of the environment. Human health and welfare, food security, industrial development and the ecosystems on which they depend, are all at risk unless water and land resources are managed more effectively in the present decade and beyond than they have been in the past.

We need to take heed of this declaration. Scarcity of freshwater may be a cause for concern in parts of Ireland and some aquatic systems which constitute a major resource from the economic, aesthetic and scientific viewpoint continue to be misused. The erroneous perception in Ireland up to the 1960s was that rivers or lakes could assimilate our wastes—if there was sufficient water the environment would take care of our wastes—and the understanding then was that 'the solution to pollution was dilution' (Mullally 1992). This is now obviously not the case. Participants at the Dublin ICWE meeting called for fundamental and new approaches for assessment, development and management of freshwater resources to be brought about 'through political commitment and involvement from the highest level of government to the smallest communities'. Our freshwater resources must be conserved and managed properly, to prevent pollution and to ensure that the next generation inherits an unspoilt resource.

ACKNOWLEDGEMENTS

The assistance of K. Clabby and J. Bowman, Environmental Research Unit, Dublin, in providing information on current lake and river water quality in Ireland in advance of publication is gratefully acknowledged.

REFERENCES

Anon. 1971 Convention on Wetlands of International Importance, Ramsar, Iran.

Anon. 1992 ICWE (International Conference on Water and the Environment, Dublin) Development Issues for the 21st Century: the Dublin statement and report of the conference. Geneva. World Meteorological Organisation.

Ashe, P. 1987 *Lasiodiamesa sphagnicola, Cardiocladius capucinus* and *Orthocladius (Eudactylocladius) fuscimanus* (Diptera: Chironomidae) new to Ireland. *Irish Naturalists' Journal* 22, 194–5.

Bowman, J. 1991 *Acid sensitive surface waters in Ireland*. Dublin. Environmental Research Unit.

Bracken, J.J., Horkan, J.P., O'Kane P. and Wise, E.J. 1977 Trophic status of Lough Leane. In W.K. Downey and G. Ní Uid (eds) *Lake pollution prevention by eutrophication control: Proceedings of a National Science Council Seminar,* 50–64. Dublin. Stationery Office.

Bradshaw, R. 1987 Erosion of mountain blanket bog. In C. O'Connell (ed.), *The IPCC guide to Irish peatlands*, 56. Dublin. Irish Peatlands Conservation Council.

Casey, T. and Dowley, A. 1992 The environmental implications for the Irish Republic of the EC Directive on urban wastewater treatment. In J. Feehan (ed.), *Environment and development in Ireland,* 424–34. The Environmental Institute, University College Dublin.

Champ, W.S.T. 1977 Trophic status of fishery lakes. In W.K. Downey and G. Ní Uid (eds) *Lake pollution prevention by eutrophication control: Proceedings of a National Science Council Seminar*, 65–78. Dublin. Stationery Office.

Clabby, K.J., Lucey, J., McGarrigle, M.L., Bowman, J.J. and Toner, P. 1992 *Water quality in Ireland, 1987–1990.* Part 1. *General assessment.* Dublin. Environmental Research Unit.

Daly, D. and Daly, E.P. 1984 A review of nitrate in groundwater and the situation in Ireland. *Irish Journal of Environmental Science* **3**, 1–13.

Daly, D. 1992 Ground water resources, a review of development, quality and pollution issues. In J. Feehan (ed.), *Environment and development in Ireland,* 476–82. The Environmental Institute, University College Dublin.

Dowling, C. and Murray D.A. 1981 The distribution of the Chironomidae (Diptera) in two Irish blanket bogs. *Proceedings of the Royal Irish Academy* 81B, 53–61.

Doyle, G. 1990 Phytosociology of Atlantic blanket bog complexes in north-west Mayo. In G. Doyle (ed.), *Ecology and conservation of Irish peatlands,* 75–90. Dublin. Royal Irish Academy.

Doyle, G. 1992 Progress and problems in the conservation of Irish peatlands, 1983 to 1991. In J. Feehan (ed.), *Environment and development in Ireland,* 511–19. The Environmental Institute, University College Dublin.

Duggan, P. and Champ, T. 1992 Lough Sheelin reviewed. In J. Feehan (ed.), *Environment and development in Ireland,* 487–98. The Environmental Institute, University College Dublin.

Duigan, C.A. 1992 The ecology and distribution of the littoral freshwater Chydoridae (Branchiopoda, Anomopoda) of Ireland, with taxonomic comments on some species. *Hydrobiologia* 241, 1–70.

Duigan, C.A. and Frey, D.G. 1987 *Eurycercus glacialis,* a chydorid cladoceran new to Ireland. *Irish Naturalists' Journal* 22, 180–3.

ECOTEC 1992 *Research and technological development for the supply and use of freshwater resources,* Part 1. *Water resources, trends and issues.* Overview report, SAST Project No 6. Brussels. Commission of the European Communities.

European Community 1985 Council directive on the assessment of certain public and private projects on the environment, 80/337/EC. Brussels. *Official Journal of the European Communities.*

European Community 1991 Council directive concerning the protection of waters against pollution caused by nitrates from agricultural sources, 91/676/EC. Brussels. *Official Journal of the European Communities.*

Illies, J. 1974 *Introduction to zoogeography.* London. Macmillan.

Kennedy, G. 1980 Drainage related changes in salmonid and invertebrate populations in the River Camowen. *Proceedings of the Institute of Fisheries Management Conference, Coleraine,* 63–8. Magee University College.

Kennan, P. S. 1992 The geological endowment. In J. Feehan (ed.), *Environment and development in Ireland,* 520–4. The Environmental Institute, University College Dublin.

Lynch, J.M. and Murray, D.A. 1994 Fishery rehabilitation and habitat enhancement following arterial drainage in Ireland. *Verhandlungen der Internationale Vereinigung für Theoretische und Angewandte Limnologie* 25, 1502–8.

McCarthy, D.T. 1977 The effects of drainage on the Trimblestown River. Benthic invertebrates and flora. *Irish Fisheries Investigations* A 16, 1–16.

McCarthy, D.T. 1983 The effects of drainage on fish stocks in the Trimblestown River. In C. Moriarty (ed), Advances in fish biology in Ireland. *Irish Fisheries Investigations* A 23, 16–19.

McCumiskey, L.M. 1991 *Water in Ireland—a review of water resources, water supplies and sewerage services.* Dublin. Environmental Research Unit.

McGarrigle, M., Lucey, J. and Clabby, K. 1992 Biological assessment of river water quality in Ireland. In P.J. Newman, M.A. Piavaux and R.A. Sweeting (eds), *River water quality—ecological assessment and control,* 371–93. Brussels. Commission of the European Communities.

Mullally, P.A. 1992 Environment and rural development at local level. In J. Feehan (ed.), *Environment and development in Ireland,* 21–7. The Environmental Institute, University College Dublin.

Neill, M. 1989 Nitrate concentrations in river waters in the southeast of Ireland and their relationship with agricultural practice. *Water Research* 23, 1339–55.

O'Grady, M.F. 1990 Rehabilitation of the Boyne. *The Engineers' Journal,* March 1990, 22–4.

O'Grady, M.F. 1991 Ecological changes over 21 years caused by drainage of a salmonid stream, the Trimblestown. *Irish Fisheries Investigations* A 33, 1–10.

O'Grady, M.F. and King, J.J. 1992 Ecological changes over 30 years caused by drainage of a salmonid stream, the Bunree River. *Irish Fisheries Investigations* A 34, 1–16.

Reynolds, J.D. 1986a Some invertebrates from Lough Roe, Co. Offaly, a rare and endangered bogland habitat. *Bulletin of the Irish Biogeographical Society* 9, 41–5.

Reynolds, J.D. 1986b Cladocera from Burren turloughs at Mullagh More, Co. Clare. *Bulletin of the Irish Biogeographical Society* 9, 51–4.

Reynolds, J.D., Cooke, D., Simon, N., Allott, N. and Brennan, N. (1994) Macroinvertebrate communities of poorly buffered afforested and non-afforested streams in Western Ireland. *Verhandlungen der Internationalen Vereinigung für Theoretische und Angewandte Limnologie* 25, 1496–1501.

Salmon Research Agency 1990 Declining sea trout stocks in the Galway/South Mayo region—a scientific appraisal. Internal report prepared for the Sea Trout Action Group. Newport, Co. Mayo. Salmon Research Agency.

Sweeting, R., Lowson, D., Hale, P. and Wright, J. 1992 Biological assessment of rivers in the U.K. In P.J. Newman, M.A. Piavaux and R.A. Sweeting (eds), *River water quality—ecological assessment and control,* 319–26. Brussels. Commission of the European Communities.

Sherwood, M. 1992 *Weather, soils and pollution from agriculture.* Dublin. AGMET, Meteorological Service.

Shiklomanova, I.A. 1990 Global water resources. *Nature and Resources* 26, 34–43.

Toner, E., O'Riordan, A. and Twomey, E. 1965 The effects of arterial drainage works on the salmon stocks of a tributary of the Moy. *Irish Fisheries Investigations* A1, 36–55

Thorn, R.H. 1986 Factors affecting the leaching of nitrate to groundwater in the Republic of Ireland. *Irish Geography* 19, 23–32.

Thorn, R. and Coxon, C. 1992 Nitrates, groundwater and the Nitrate Directive. In J. Feehan (ed.), *Environment and development in Ireland,* 483–6. The Environmental Institute, University College Dublin.

Ward, J.V. 1989 The four dimensional nature of lotic ecosystems. *Journal of the North American Benthological Society* 8, 2–8.

Wood, B. 1989 *Lough Neagh—managing to stay resourceful.* 1988 Went Memorial Lecture, Royal Dublin Society.

Wood, B. and Gibson, C.E. 1973 Eutrophication and Lough Neagh. *Water Research* 7, 173–87.

In: J.D. Reynolds (ed.) 1996 *The conservation of aquatic systems*, 20–30. Dublin. Royal Irish Academy.

LOWER SHANNON BASIN—ATHLONE TO KILLALOE

J. Bowman

Environmental Protection Agency,
Dun Laoghaire, Co. Dublin

ABSTRACT

The substantial growths of planktonic algae measured in Lough Derg during 1991–2 (median chlorophyll 15–18mg/m^3) show significant increases when compared with the previous values measured in the 1976–8 period (median chlorophyll 7–12mg/m^3) and indicate that the lake is now strongly eutrophic. A marked decrease in water transparency and increase in phosphorus in the lake were also recorded. These changes were greatest in the middle and lower sections. High phosphorus concentrations were measured at several locations in the inflowing rivers, particularly the rivers Brosna, Little Brosna, Nenagh and, to a lesser extent, Graney and Suck. Significant increases in the inputs to Lough Derg of molybdate reactive phosphate in the Shannon, Little Brosna and Nenagh are apparent when a comparison is made between the 1976–8 and the 1987–91 nutrient loadings. Numerous recommendations are proposed to reduce these phosphorus inputs, involving the installation of phosphorus removal facilities at nine waste treatment plants and a reduction in the phosphorus losses from the agricultural sector.

INTRODUCTION

The Lower Shannon Basin, from Athlone to Killaloe, is unusual amongst Irish freshwaters in that it was the subject of a limnological investigation in the early part of this century. This study, carried out from a research laboratory located at Hayes Island near Portumna, had as its aim 'to lay down a foundation of knowledge which will serve as a basis for the future regulation, conservation and development of the freshwater fisheries'. Particular attention was paid to the planktonic organisms, the zooplankton being especially well documented. The reports of the investigations (Southern and Gardiner 1926; 1938; Southern 1935) provide useful information on the biological and, to a lesser degree, the chemical characteristics of Lough Derg and the River Shannon at Portumna as they were seventy years ago.

Work ceased at the Laboratory in 1923 and no significant study of the lake or river was carried out for the next fifty years.

Brief investigations of the water quality of the Shannon System in 1972 (Toner *et al.* 1973) and lake only in 1973–4 (Flanagan and Toner 1975) were carried out, the results of which suggested that the lake was becoming enriched. These findings led to a major three-year study of the water quality of the lake and its feeder streams commencing in 1976. The objective of this study was to establish a modern baseline for the water quality of Lough Derg, paying particular attention to nutrient and plankton concentrations. An estimate of the nutrient input to the lake was made. This study concluded that the trophic status of Lough Derg was transitional between mesotrophic and eutrophic (Bowman 1985). Further brief investigations in 1982, 1984 (Bowman 1985) and 1986 (ERU unpublished data) showed a marked deterioration in water quality when compared with that measured in 1976–8 and a substantial increase in phytoplankton growth.

The recent deterioration in the condition of Lough Derg in conjunction with a greater awareness of the importance of water quality to its beneficial uses led to a further detailed study of the River Shannon catchment from Athlone to Killaloe (Bowman *et al.* 1993). The purpose of this investigation was to identify the principal sources of nutrients in the catchment and make recommendations for improving the water quality of the lake. The requirements of this survey warranted a broader investigation of water quality in the catchment than had been carried out previously. In addition to the more important streams flowing directly to Lough Derg, the principal influents of the River Shannon, south of Athlone, were examined over their length. The river data cover the period 1987–92 and are drawn principally from the water quality programmes of the Regional Water Laboratories at Castlebar and Kilkenny, supplemented by data from the ERU Laboratory in Dublin. The study of the lake covers the period 1991–2 and was carried out by the ERU Laboratory.

BIOLOGICAL CHARACTERISTICS

Several aspects of the biological properties of the Lower Shannon Basin have been examined, principally to assess water quality conditions in rivers and lakes. The plankton studies have been largely confined to investigations of Lough Derg, whereas extensive inspections of the macroinvertebrate communities in the rivers have been carried out on a regular basis over the last 20 years as part of the biological monitoring programme of An Foras Forbartha/ERU.

Phytoplankton of the River Shannon

A detailed study of the phytoplankton in the River Shannon was carried out during the 1922–3 investigation only. Diatoms were the main components of the riverine phytoplankton standing crop in the spring with *Fragilaria* spp and *Melosira* spp prominent. These forms were co-dominant during the summer months with the green algal species *Ankistrodesmus falcatus* and the flagellates *Mallomonas* spp and *Dinobryon* spp. The cyanobacteria *Oscillatoria* spp assumed dominance in the autumn months and remained prominent through the winter.

Phytoplankton of Lough Derg

The pattern of phytoplankton succession in Lough Derg has not changed significantly during the last 70 years. The overwinter populations are regularly dominated by mixed cyanobacteria and diatom populations. With the onset of increasing light and temperature the diatoms become more prominent. Frequently one or more of the pennate forms *Diatoma elongatum, Asterionella formosa, Tabellaria flocculosa, Synedra* spp and *Fragilaria* spp are most conspicuous in the early spring. By late March centric diatoms, particularly *Melosira* spp and *Stephanodiscus* spp, are the main components of the springtime maxima in plankton growth. These forms persist through the rest of the spring and early summer months when, in conjunction with the emerging green algae, they may be a prominent element in reduced standing crops.

The phytoplankton of Lough Derg during the summer months is characterised by the prominence of the dinophycean *Ceratium hirundinella* and the chrysophytan *Dinobryon* spp in diverse populations. In recent years *Melosira granulata* and the colonial cyanobacteria have become more prominent, with the former frequently the dominant form. The cyanobacteria, while ever present and quantitatively important in the lake, normally do not assume dominance until the autumn. The filamentous *Oscillatoria* spp frequently represent in excess of 50% of the biomass at this time. Thereafter a quantitative decline in the phytoplankton standing crop occurs and a return to a more balanced, though less diverse, population is noted.

Marked changes in the abundance of several of the principal planktonic algae in Lough Derg are inferred from estimates of the cell biomass during the various examinations that have been carried out. Significant increases in the size of many of the diatom and cyanobacteria populations have occurred; however, little change in the representation of the green algae is noted. Overall the phytoplankton standing crop in the lake, as indicated by measurements of the algal pigment chlorophyll (see Table 2) and by estimates of cell biomass, has increased dramatically since 1922, but most noticeably in the last 10–15 years. The extent of algal development in the lake during the early part of the century was consistent with that of an oligotrophic system. The corresponding data recorded in the 1976–8 study suggested a transitional state between a mesotrophic and slightly eutrophic condition. A substantial increase in the phytoplankton standing crop was recorded in the subsequent examinations commensurate with the lake being strongly eutrophic.

Zooplankton of Lough Derg

Two investigations of the zooplankton populations in Lough Derg have been carried out. The first, during 1921–2, was confined to the upper portion of the lake in the vicinity of the present station E (Fig. 1), while the second, in 1976–8, covered the whole lake. Because of different sampling techniques care is needed in comparing the results of the two periods quantitatively.

No significant change in the composition of the zooplankton population of Lough Derg was noted over the intervening half century. The numerically dominant taxa were few and much the same in each period; the majority of the taxa were minor components of the population and occurred infrequently or in small numbers. The most notable change between the two sets of data appears to have occurred in the increased abundance of the non-diaptomid copepods (mainly Cyclopoida). This increase in the cyclopoid population may reflect the increased phytoplankton growth in the lake since the earlier study; however, other changes frequently associated with increased trophic

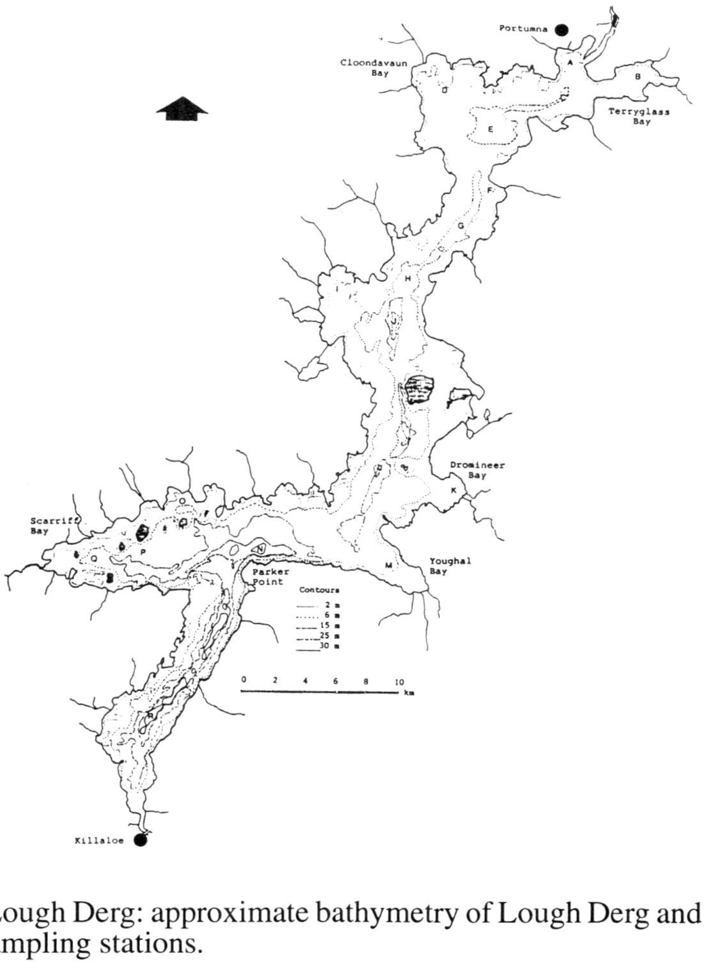

Fig. 1—Lough Derg: approximate bathymetry of Lough Derg and locations of lake sampling stations.

status were not apparent at this time, suggesting that the changes in lake productivity since 1923 were limited. An increase in the number of *Diaptomus* was also apparent, whereas that of *Bosmina* had decreased. Little change was noted in the abundance of *Daphnia* and the two predatory Cladocera.

Macroinvertebrate communities of Lough Derg

Limited data exist on the macroinvertebrate communities of Lough Derg. The dominant organisms in the littoral and sublittoral areas of the lake since the 1920s are *Asellus* spp. Fewer Mollusca, Ephemeroptera, Oligochaeta, Chironomidae, Hirudinea and Tricladida were also recorded. *Ephemera danica* has been a constant component of the bottom fauna of the lake throughout the years despite some concern about the impact of peat silt on this species. The sublittoral fauna is composed mainly of chironomids and Tubificidae.

Macroinvertebrate communities of rivers

Diverse populations, containing good representation of each of the main macroinvertebrate groups, have been consistently recorded at locations in the catchment. Many different river types are represented, including acidic headwaters, fast flowing first-, second- and third-order rivers and slow flowing lowland rivers. Good examples of fauna, typical of these river types, are present in the catchment. Outstanding in this respect is the River Camcor, which exhibits a rich diversity of organisms (K.J. Clabby, pers. comm.). However, a decline in water quality conditions at some locations has led to a restricted fauna, typically resulting in depletion or elimination of Plecopteran and Ephemeropteran taxa and increased representation of the Oligochaeta and Chironomidae. The River Shannon and the lower reaches of its principal tributaries are deep, slow flowing watercourses and in consequence they have a naturally restricted benthic fauna. However, good diversity is found at the edges of these rivers, notably in the River Shannon downstream of Athlone.

Fish

Fish stocks of many of the individual rivers in the lower River Shannon catchment have been studied by the Central Fisheries Board in recent years. In addition, detailed investigations of eel stocks (T.K. McCarthy, pers. comm.) and salmon (M.J. O'Farrell, pers. comm.) are nearing completion. Naturally spawned juvenile salmon were well distributed in the catchment and substantial salmonid stocks were reported in all waters examined; however, major fish kills, attributable to the effects of effluent from agricultural activities, were reported in sections of these rivers (M.F. O'Grady, pers. comm.).

Variable catches of brown trout (*Salmo trutta* L.) have been reported in the lower section of Lough Derg in recent years; however, returns from the remainder of the lake indicate poor game angling (P. Bourke, pers. comm.). The lower section of the lake in the area north of Killaloe is also an excellent coarse fishery noted for its bream. Good game fishing is reported in the rivers Little Brosna and Nenagh and to a lesser extent in the River Brosna. However, a decline in the River Suck fishery is reported in recent years.

WATER CHEMISTRY

Lough Derg

Chemical observations in the 1922–3 survey were confined to the River Shannon at Portumna and the upper section of the lake at locations corresponding to stations A, B, D and E of the more recent surveys (Fig. 1), which have covered the entire lake.

Marked fluctuations in the ionic content of the lake water have been recorded, with the greatest variation and highest concentrations generally occurring in the upper section of the lake, reflecting the influence of the River Shannon. Lesser concentrations were noted in the middle and lower sections and minimum values were recorded during the summer months when loads carried by the inflowing streams were at their lowest.

The lake waters were alkaline in reaction on all occasions and show little variation since 1922; pH values were in excess of 8.0 on most sampling days. Minimum values occurred during the winter and were frequently associated with periods of high run-off; maximum values coincided with peaks of phytoplankton development and photosynthesis. The variations in water colour showed strong seasonal trends in all sections of the lake with high values (>50 Hazen) recorded during the winter. It is probable that the greater part of the material imparting this colour is in the form of suspended and dissolved humic material derived from the bogland in the River Shannon catchment. No measurements of colour were made in the course of the 1922–3 investigation; however, 'brown floods' were observed in the River Shannon when heavy rainfall followed drought and resulted in considerable turbidity in the upper section of the lake. It was inferred that the staining of the water, at that time, was due to silt brought in by the rivers Little Brosna and Suck.

An investigation of the release of phosphate from humic substances under the influence of UV irradiation was carried out in 1991–2 (Bowman *et al.* 1993). The results suggest that the impact of phosphorus from this source on the phosphate budget of the River Shannon and Lough Derg is quite small. Apart from occasional periods when a significant phosphate release may occur, the overall balance appears negative. Humic substances from the bogs may, however, enhance the impact of other phosphate sources by acting as a transporter of phosphate.

A progressive decline in water transparency in Lough Derg has been measured over the last 70 years (Table 1). This was most apparent over the last decade in the middle and lower sections of the lake, where a reduction of 0.5m in the median value was recorded. Transparency showed a strong negative correlation with colour in the upper section of Lough Derg and less strongly with algal development elsewhere in the lake.

Table 1—Comparison of minimum, median and maximum values for water transparency (m) at representative stations in the upper, middle and lower sections of Lough Derg as measured in the investigations since 1922.

	Station E			*Station J*			*Station N*		
	Min.	*Med.*	*Max.*	*Min.*	*Med.*	*Max.*	*Min.*	*Med.*	*Max.*
1922	1.4	2.5	4.0			4.0			
1972*			2.8			4.0			
1976	0.9	1.7	2.5	1.0	2.2	3.2	1.6	2.6	3.2
1977	1.5	2.1	2.5	1.8	2.2	3.0	2.2	2.5	3.4
1978	1.0	1.8	2.4	1.1	2.1	3.1	1.4	2.3	4.2
1976–8	0.9	1.8	2.5	1.0	2.1	3.2	1.6	2.5	4.2
1982*			1.3			1.5			1.8
1984**	1.5	2.0	2.4	1.4	2.1	2.8	2.8	2.9	4.0
1986**	0.5	1.7	2.4	0.6	1.9	2.2	0.7	2.1	3.1
1991–2	0.9	1.6	1.8	1.2	1.6	2.6	1.4	2.0	2.6

* One set of samples only; ** summer samples only

Satisfactory levels of dissolved oxygen were measured in Lough Derg on most occasions; however, severe oxygen depletion (<20% saturation) was measured in the hypolimnion at the deeper stations associated with thermal stratification during the summer months in most recent years.

Nutrient levels in Lough Derg have been measured over the last 20 years only. A significant increase in the levels of molybdate reactive phosphate (MRP) and total phosphorus is indicated when the relatively low concentrations recorded in the later period of the 1977–8 survey are compared with those measured over the 1991–2 period (Table 2). No significant change in the median concentrations of oxidised nitrogen and ammonium were indicated between the periods of examination; however, the data indicated a declining trend for the latter parameter at station N in the lower section of the lake.

Considerable variation in the mean and median silica concentrations was recorded in the lake and a declining trend also was indicated for this parameter at station N, possibly reflecting uptake by the increased diatom population.

Table 2—Comparison of the median values for MRP, total phosphorus, oxidised nitrogen, ammonium, silica and chlorophyll at representative stations in the upper, middle and lower sections of Lough Derg during the periods (a) June 1977 to October 1978 and (b) June 1991 to October 1992.

	MRP	*Total-P*	*Ox.N*	*NH_4*	*Silica*	*Chlorophyll*
Station E						
(a)	5	15	840	15	1.15	11.8
(b)	11	48	980	17	1.07	15.3
Station J						
(a)	3	20	740	15	0.85	10.0
(b)	6	45	760	20	0.72	16.8
Station N						
(a)	3	16	895	15	1.00	7.6
(b)	5	43	760	10	0.75	18.2

Rivers

Instances of acute and chronic organic pollution and nutrient enrichment were recorded in some of the rivers and streams in the River Shannon catchment (Fig. 2). Markedly increased concentrations of MRP and, where measured, total phosphorus were recorded downstream of the municipal and industrial effluent discharge points on many of these rivers. The elevated phosphorus concentrations persisted, in many cases, over the remaining reaches of the rivers. In contrast, concentrations of oxidised nitrogen (NO_2 + NO_3) increased gradually over their courses. The worst affected reaches were the River Brosna downstream of its confluence with the Moate Stream and the River Clodiagh, the River Little Brosna downstream of the confluence with the River Bunow and the River Nenagh downstream of the three main effluent discharges at Nenagh. Moderate degrees of enrichment were recorded in the lower reaches of the rivers Suck and Graney. Increased, but still moderate, phosphorus levels were also recorded in the River Shannon downstream of Athlone and in the reach from Banagher to Portumna, the latter reflecting the influence of the confluence with the rivers Brosna and Little Brosna. Substantial phytoplankton populations were noted in the river at all sampling points, particularly in the lower reaches.

Trends in nutrient concentrations in these rivers suggest significant increases in the levels of MRP in recent years, particularly in the rivers Shannon, Little Brosna and Nenagh. Slight increases in ammonium concentrations were measured in the rivers Little Brosna and Nenagh; however, elsewhere no change or reduced ammonium values were recorded. With the exception of the rivers Cappagh and Killimor in east Galway, increased concentrations of oxidised nitrogen were measured in the rivers.

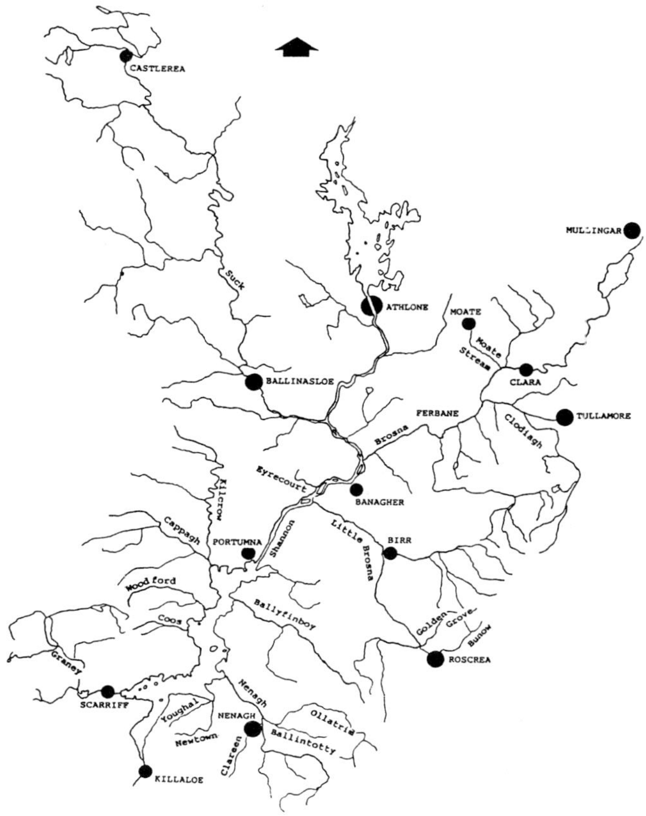

Fig. 2—Lower Shannon catchment—Athlone to Killaloe: principal towns and rivers.

NUTRIENT INPUTS

The estimated total annual inputs of MRP and dissolved inorganic nitrogen (DIN) = nitrate + nitrite + ammonium (NO_3 + NO_2 + NH_4) to Lough Derg for the period 1987–91, derived from the inflowing rivers and streams, ranged from 146 to 204 tonnes P and 5977 to 8300 tonnes N respectively (Bowman *et al.* 1993). The River Shannon, representing 85% of the catchment area, was by far the largest contributor, accounting for 70–76% of the MRP and 70–80% of the DIN loads. The River Nenagh, representing 3% of the catchment area, was the next highest contributor of nutrients, accounting for 13–16% of the MRP and 4–5% of the DIN loads. The remaining rivers had less significant inputs. These estimated loads represent marked increases over those estimated for the 1976–8 period (Bowman 1985), particularly in the case of MRP.

CONCLUSIONS

The results of investigations of Lough Derg and its catchment since 1922 indicate increased enrichment of the lake and a progressive deterioration in the water quality of the lake and inflowing rivers, particularly over the last 15 years. The most striking differences in water quality of the lake are the changes recorded in both planktonic algal growth and water transparency. The estimated extent of algal development in the lake in 1922–3 suggested a relatively unenriched system; however, the substantial developments of planktonic algae measured in recent years are consistent with those of a highly eutrophic system.

The enriched condition of Lough Derg is attributed to nutrients, particularly phosphorus, arising from discharges in the lake's extensive catchment. High concentrations of MRP (phosphate) were recorded in a number of rivers flowing into the River Shannon between Athlone and Portumna, although concentrations in the River Shannon itself remained at a moderate level. Of the smaller rivers flowing into the lake the River Nenagh and to a lesser extent the River Graney were consistently enriched. The significant increase in the total phosphorus and MRP concentrations in these rivers, measured over the last 17 years, has resulted in the substantially increased algal crops in Lough Derg that are adversely affecting its water quality.

Notwithstanding the apparent stabilisation of the rate of enrichment of Lough Derg in the last ten years, as suggested by the levels of planktonic algal growth, the water quality of the lake is now in a highly eutrophic condition. It is clear that the phosphorus inputs to the lake substantially exceed the natural or background levels that sustain normal plant growth. Should further increased inputs of phosphorus occur, the resulting additional deterioration in the water quality of the lake will have an adverse effect at all levels of the biota, resulting in major changes in the structures of the aquatic ecosystems.

To prevent further deterioration in the water quality of Lough Derg and to reduce algal development to an acceptable level in respect of the beneficial uses of the lake an integrated approach to the control of phosphorus discharges in the catchment is proposed (Bowman *et al.* 1993). This will involve reducing significantly the phosphorus content of the waste discharges from the municipal, industrial, public, agricultural and domestic sectors.

It is proposed that phosphorus removal facilities be installed at the sewage treatment plants at Nenagh, Roscrea, Tullamore, Athlone, Ballinasloe, Birr, Moate and Portumna and at an industrial plant in Nenagh. The use of phosphate-free detergents is encouraged with particular attention paid to recommending their use in lakeside dwellings, in the main urban areas and for milk-churn cleaning on farms. A wide range of recommendations is proposed to reduce phosphorus losses to water from the agriculture sector. These are principally concerned with the spreading of natural and artificial fertilisers and the disposal of animal manure wastes on land. The completion of the installation of onshore pump-out facilities, including sluice tanks, at harbours on the River Shannon and Lough Derg to receive waste from cruisers is recommended. All new cruisers are to have waste-holding tanks and standard pump-out equipment.

It is estimated that the decreased inputs of phosphorus to Lough Derg resulting from these measures will reduce the amounts of algal and rooted plant growth in the lake to the levels recorded 15 years ago. The lake was categorised at that time as being in a transitional state between mesotrophic and eutrophic, and the water quality was regarded as being in a satisfactory condition. A reduction in the phosphorus input to the lake to restrict algal growths to an oligotrophic or unenriched level may not be desirable or achievable. As a renowned game fishery Lough Derg requires a moderate level of production to generate sufficient food organisms and this capacity could be impaired by the level of production associated with oligotrophic status. Furthermore to achieve a level of algal development in Lough Derg consistent with an oligotrophic status would require a reduction in the phosphorus input to the background or natural run-off level for the catchment and hence may not be a realistic target.

REFERENCES

BOWMAN, J.J. 1985 *The trophic status of Irish lakes: Lough Derg.* Dublin. An Foras Forbartha.

BOWMAN, J.J., McGARRIGLE, M.L. and CLABBY, K.J. 1993 *Lough Derg: an investigation of eutrophication and its causes—Part 1. Water quality assessments, nutrient sources, conclusions and recommendations. A report to the Lough Derg Working Group.* Dublin. Environmental Research Unit.

FLANAGAN, P.J. and TONER, P.F. 1975 *A preliminary survey of Irish lakes.* Dublin. An Foras Forbartha.

SOUTHERN, R. 1935 Reports from the Limnological Laboratory—III. The food and growth of brown trout from Lough Derg and the River Shannon. *Proceedings of the Royal Irish Academy* 42B, 87–112.

SOUTHERN, R. and GARDINER, A.C. 1926 Reports from the Limnological Laboratory—I. The seasonal distribution of the Crustacea of the plankton in Lough Derg and the River Shannon. *Fisheries, Ireland, Scientific Investigations*, 1926, I.

SOUTHERN, R. and GARDINER, A.C. 1938 Reports from the Limnological Laboratory— IV. The phytoplankton of the River Shannon and Lough Derg. *Proceedings of the Royal Irish Academy* 45B, 89–124.

TONER, P.F., CLABBY, K.J. and BOWMAN, J.J. 1973 *Report on results of a biological and chemical survey of the main river channel and lakes of the Shannon system, between Battle Bridge and Killaloe, undertaken in July 1972.* Dublin. An Foras Forbartha.

In: J.D. Reynolds (ed.) 1996 *The conservation of aquatic systems*, 31–37. Dublin. Royal Irish Academy.

RIVER WATER QUALITY IN IRELAND 1987–90

M.L. McGarrigle,[1] J. Lucey[2] and K.C. Clabby[3]

Environmental Protection Agency

ABSTRACT

The national survey of rivers in Ireland used the Quality Rating System—a biological index system—for assessing water quality. In the period 1987–90 over 4200 sites were surveyed on 13,470km of river channel. Of this length, 76.4% was unpolluted, 12.1% slightly polluted, 10.3% moderately polluted and 1.2% seriously polluted. The main trends observed since the surveys began in 1971 are an increase in the length of slightly and moderately polluted river channel and a decrease in the length of seriously polluted stretches. The control of eutrophication in rivers is seen as the main task currently facing Irish water quality managers.

INTRODUCTION

A single scheme of biological river quality classification, the Quality Rating System, is used in national surveys of Irish rivers. This biotic index was first described by Flanagan and Toner (1972) and has been in use in Ireland since 1971. Minor revisions to the Quality Rating System were outlined by Clabby and Bowman (1979), Clabby (1981) and Clabby *et al.* (1992). The scheme is based primarily on the macroinvertebrate fauna, but it also incorporates observations on algal abundance and other site-specific environmental factors (McGarrigle *et al.* 1992).

The biological programme is currently carried out by a team of three biologists. The complete national survey takes four years to complete but the most seriously polluted locations are surveyed annually, where possible. The biological survey has been the primary means for assessing geographical and temporal trends in river water quality in Ireland since 1971. A review of river water quality for the period 1971–89 was given in Clabby (1981).

[1] EPA, Regional Inspectorate, Castlebar, Co. Mayo.
[2] EPA, Regional Inspectorate, Kilkenny, Co. Kilkenny.
[3] EPA, Regional Inspectorate, Pottery Road, Dun Laoghaire, Co. Dublin.

Interim annual or biennial reports have been produced since then (Clabby *et al.* 1982 to 1994). Major national reviews of water quality (including rivers, lakes and estuaries) are undertaken at regular intervals, assessing national water quality data gathered over a four-year period (e.g. WPAC 1983; Toner *et al.* 1986; Clabby *et al.* 1992).

RESULTS

In the period 1987–90, some 4550 assessments were made on 4239 sites on Irish rivers, covering a total surveyed river channel length of 13,470km. The river water quality map illustrates these results on a site-by-site basis (Clabby *et al.* 1991a). Individual sites are colour-coded for water quality status. Approximately 4000 individual sample stations on over 1000 rivers are illustrated on this map. The overall breakdown of channel length is indicated in Table 1, together with the colour-coding scheme used.

Table 1—River water quality in Ireland 1987–90.

Category	*%*	*km*	*Map colour code*
Unpolluted	76.4	10,291	Blue
Slightly polluted	12.1	1,630	Green
Moderately polluted	10.3	1,387	Orange/yellow
Seriously polluted	1.2	162	Red

There has been a reduction in the extent of serious pollution between 1971 and 1990. During the same period, however, the length of channel suffering from slight and moderate pollution has increased significantly. The main feature of the long-term survey is an increase in the level of eutrophication in rivers.

Changes in the initial 2900km baseline, which was first surveyed in 1971, are illustrated in Fig. 1. A more extensive 6900km baseline was surveyed in the periods 1982–6 and 1987–90 (Fig. 2). Figure 3 compares the 6900km baseline with the full 13,470km survey results. The trends of decreasing serious pollution and of increasing moderate and slight pollution are particularly evident in the 2900km baseline, but these trends are also apparent in the 6900km baseline.

CAUSES OF POLLUTION

The causes of serious pollution are shown in Table 2. Most cases of serious pollution are due to chronic long-term discharges from overloaded or inadequate treatment plants—town sewage or industrial—particularly from older, outdated plants. The chronic nature of many instances of serious pollution is reflected in the fact that they are reported on annually in the interim biological reports on river quality (Clabby *et al.* 1982 to 1994).

Table 2—Causes of serious river pollution 1987–90 by channel length and percentage breakdown.*

	km	%
Sewage treatment plant effluent	37.7	24.0
Industrial effluent	39.3	25.0
Combination of sewage and industrial	24.0	15.5
Agricultural wastes	35.8	23.0
Unknown and peat silt	19.3	12.5

*This does not include a further 8km of seriously polluted stream length not shown on the OS catchment map.

Slightly and moderately polluted river stretches are frequently found downstream of sewage treatment works that do not have a phosphate removal stage. Installation of improved secondary treatment plants has resulted in improvements in some of the seriously polluted channel, such that it now falls into the moderately or slightly polluted categories. A long-term increase in the levels of diffuse runoff from land is also an important factor in the decline in river quality. Diffuse sources are difficult to pinpoint and quantify and thus it is more difficult to attribute pollution to particular sectors in such cases. It is hoped, however, to include a detailed analysis of the causes of slight and moderately polluted river channel in the national review to be published by the EPA, which will cover the period 1991–4.

The increase in the level of slight and moderate pollution is undoubtedly the most pressing water quality problem currently facing water quality managers. It is believed that much of this deterioration is due to eutrophication caused by above-background phosphate levels. Indeed biological quality and phosphate levels were clearly correlated in a survey encompassing over 1200 river sites in Ireland (Clabby *et al.* 1992). Eutrophication becomes evident at sites that have annual median orthophosphate levels greater than 30µg P l^{-1}. As Irish rivers are generally classified as salmonid waters, strict control of eutrophication is necessary in order to protect trout and salmon stocks. These oxygen-sensitive fish are at risk particularly during warm, dry summers when fish kills due to low night-time dissolved oxygen levels are most likely. The EU Freshwater Fish Directive (78/659/EEC) does not set phosphate standards for flowing waters and there are currently no Irish national standards for phosphate levels in rivers (McGarrigle 1993). Water quality management plans have specified phosphate limits for certain catchments in the past, but it now appears that significantly lower phosphate standards will be required in order to control river eutrophication.

Management of diffuse runoff from agricultural land is likely to become a key area in improving river water quality. A detailed study of streams in the Lough Conn catchment has demonstrated a change in runoff patterns in the period 1982–90 (McGarrigle *et al.* 1993). Recent data show an increase in phosphate runoff during wet weather, whereas there was no correlation between rainfall and phosphate during the period 1979–82. It is believed that a similar situation pertains in the Lee Valley (V. Power, pers. comm.). The practice of building up soil fertility and of 'insurance' levels of phosphate (i.e. more than that required by crops in the next year) may be in part responsible for the increase in diffuse runoff from agricultural lands (Tunney 1990).

Details of water quality in rivers, lakes and estuaries are available in a two-volume report on water quality in Ireland for the period 1987–90 (Clabby *et al.* 1992; Flanagan and Larkin 1992), and the river water quality map comes with the report (Clabby *et al.* 1991a).

REFERENCES

CLABBY, K.J. 1981 *The national survey of Irish rivers: a review of biological monitoring 1971–1979.* WR/R12. Dublin. Water Resources Division, An Foras Forbartha.

CLABBY, K.J. and BOWMAN, J.J. 1979 Report of the Irish participants. In P.F. Ghetti (ed.), *Third Technical Seminar on Biological Water Assessment Methods, Parma, 1978,* vol. 1. Commission of the European Communities.

CLABBY, K.J., LUCEY, J. and McGARRIGLE, M.L. 1982 *The national survey of Irish rivers: river quality investigations—biological. Results of the 1980 and 1981 investigations.* WR/R14. Dublin. Water Resources Division, An Forbartha.

CLABBY, K.J., LUCEY, J. and McGARRIGLE, M.L. 1984 *The national survey of Irish rivers 1982–3: biological investigations of quality in selected rivers and streams.* Dublin. Water Resources Division, An Foras Forbartha.

CLABBY, K.J., LUCEY, J. and McGARRIGLE, M.L. 1986 *The national survey of Irish rivers: biological monitoring. Interim report for 1984.* WR/R17. Dublin. Water Resources Division, An Foras Forbartha.

CLABBY, K.J., LUCEY, J. and McGARRIGLE, M.L. 1986 *The national survey of Irish rivers: biological monitoring. Interim report for 1985.* WR/R18. Dublin. Water Resources Division, An Foras Forbartha.

CLABBY, K.J., LUCEY, J. and McGARRIGLE, M.L. 1988 *The national survey of Irish rivers: biological monitoring. Interim report for 1986.* WR/R20. Dublin. Water Resources Division, An Foras Forbartha.

CLABBY, K.J., LUCEY, J. and McGARRIGLE, M.L. 1989 *Interim report on the biological survey of river quality: results of the 1987 investigations.* WR/R21. Dublin. Environmental Research Unit.

CLABBY, K.J., LUCEY, J. and McGARRIGLE, M.L. 1989 *Interim report on the biological survey of river quality: results of the 1988 investigations.* Dublin. Environmental Research Unit.

CLABBY, K.J., LUCEY, J. and McGARRIGLE, M.L. 1990 *Interim report on the biological survey of river quality: results of the 1989 investigations.* Dublin. Environmental Research Unit.

CLABBY, K.C., LUCEY, J.L. and McGARRIGLE, M.L. 1991a *Map: Ireland river quality 1987–1990.* Dublin. Environmental Research Unit.

CLABBY, K.J., LUCEY, J. and McGARRIGLE, M.L. 1991b *Interim report on the biological survey of river quality: results of the 1990 investigations*. Dublin. Environmental Research Unit.

CLABBY, K.J., LUCEY, J. and McGARRIGLE, M.L. 1993 *Interim report on the biological survey of river quality: results of the 1991 investigations*. Dublin. Environmental Research Unit.

CLABBY, K.J., LUCEY, J. and McGARRIGLE, M.L. 1993 *Interim report on the biological survey of river quality: results of the 1992 investgations.* Dublin. Environmental Research Unit.

CLABBY, K.J., LUCEY, J. and McGARRIGLE, M.L. 1994 *Interim report on the biological survey of river quality: results of the 1993 investgations.* Dublin. Environmental Research Unit.

CLABBY, K.C., LUCEY, J.L., McGARRIGLE, M.L., BOWMAN, J., FLANAGAN, P.J. and TONER, P.F. 1992 *Water quality in Ireland 1987–1990—Part 1. General assessment.* Dublin. Environmental Research Unit.

FLANAGAN, P.J. and LARKIN, P.M. (eds) 1992 *Water quality in Ireland 1987–1990—Part 2: River quality data.* Dublin. Environmental Research Unit.

FLANAGAN, P.J. and TONER, P.F. 1972 *The national survey of Irish rivers: a report on water quality.* Dublin. Water Resources Division, An Foras Forbartha.

McGARRIGLE, M.L. and LUCEY, J. 1983 Biological monitoring in freshwaters. *Irish Journal of Environmental Science* 2(2), 1–18.

McGARRIGLE, M.L., LUCEY, J.L. and CLABBY, K.C. 1992 Biological assessment of river water quality in Ireland. In P.J. Newman, M.A. Piavaux and R.A. Sweeting (eds), *River water quality, ecological assessment and control* (EUR 14606 EN-FR). Luxembourg. Commission of the European Communities.

McGARRIGLE, M.L., CHAMP, W.S.T., NORTON, R., LARKIN, P.M. and MOORE, M. 1993 *The trophic status of Lough Conn.* Mayo County Council and others.

ORDNANCE SURVEY OF IRELAND 1958 *River catchment map of Ireland*. Dublin.

TONER, P.F., CLABBY, K.J., BOWMAN, J.J. and McGARRIGLE, M.L. 1986 *Water quality in Ireland, the current position—Part one: General assessment,* WR/G15. Dublin. Water Resources Division, An Foras Forbartha.

TUNNEY, H. 1990 Phosphorus in agriculture—environmental implications. In *Environmental impact of landspreading of wastes.* (Proceedings of a seminar held at Johnstown Castle Centre for Soils and Environmental Research and Development, 30–1 May 1990),103–9.

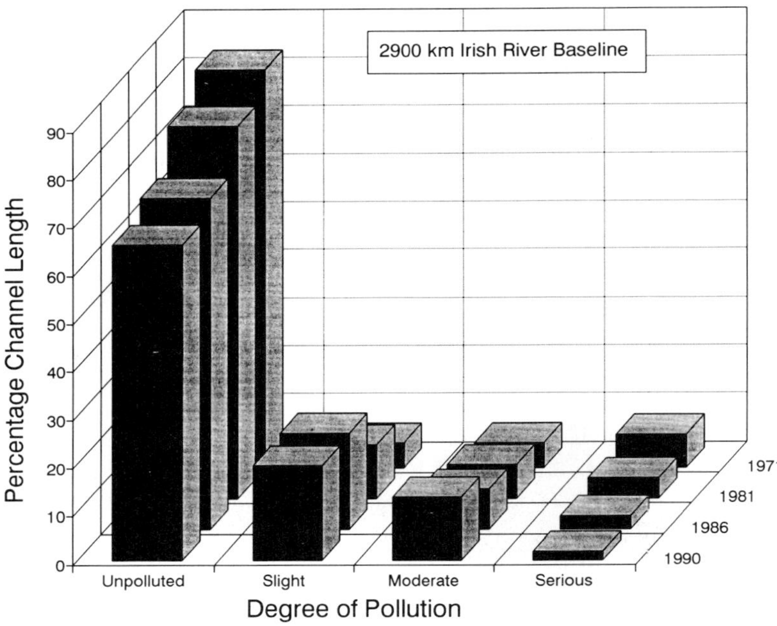

Fig. 1—Water quality in Irish rivers: trends in the 2900km baseline compared for 1971, 1981, 1986 and 1990.

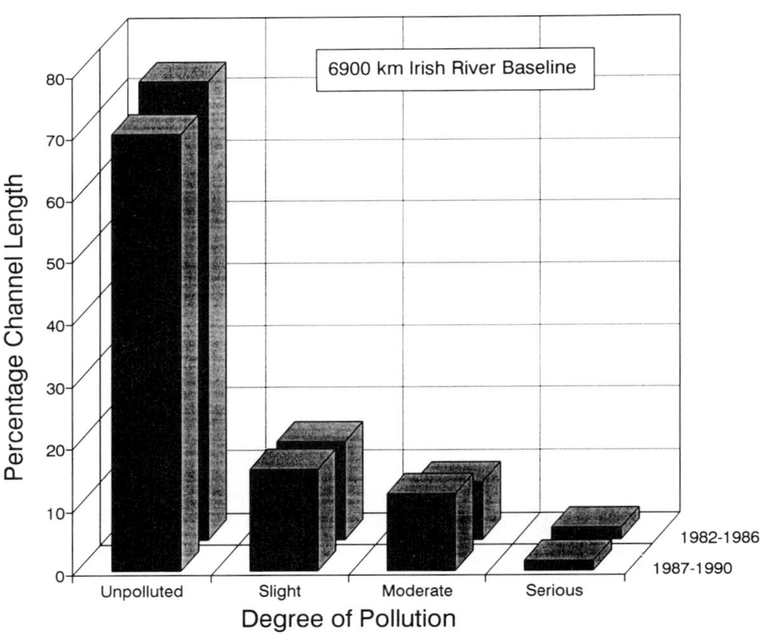

Fig. 2—Water quality in Irish rivers: trends in the 6900km baseline compared for the periods 1982–6 and 1987–90.

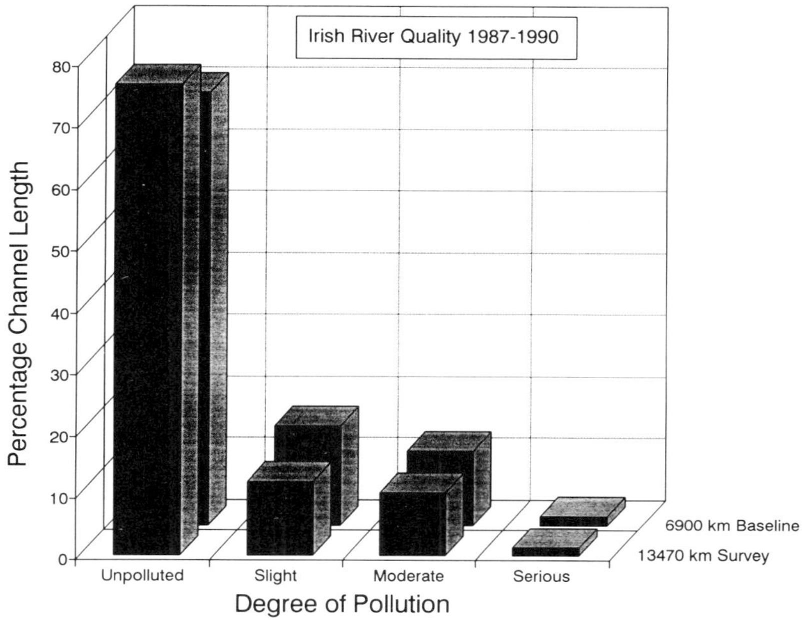

Fig. 3—Water quality in Irish rivers for the period 1987–90, comparing the 6900km baseline with the full 13,470km surveyed during the period.

In: J.D. Reynolds (ed.) 1996 *The conservation of aquatic systems*, 38–46. Dublin. Royal Irish Academy.

TURLOUGHS, THEIR SIGNIFICANCE AND POSSIBILITIES FOR CONSERVATION

Julian D. Reynolds

Department of Zoology, Trinity College, Dublin

ABSTRACT

Turloughs, from the Irish *tuarloch* (dry lake), are temporary water bodies on Carboniferous limestone, most representative of the western third of Ireland. They are distinctive in that their filling and emptying are, to some extent at least, through connections with the ground water, typically via a boulder-filler *slugaire* or swallow-hole, and are thus neither predictable nor strongly seasonal, but reflect relatively local weather patterns. These features lead to a characteristic vegetation, but sparse and unpredictable aquatic faunal communities. Turlough conservation requires sensitive management of the regional catchment, both surface and underground.

DESCRIPTION AND DISTRIBUTION

Turloughs are the characteristic temporary water-bodies of western Ireland. They are analogous in some ways to vernal pools, but distinguished from them in that (i) they fill to some extent and chiefly empty via connections with the ground water, and (ii) their changing levels are not always strongly seasonal, but often reflect relatively local weather patterns. Unlike some other freshwaters, they do not constitute ecosystems, except in a temporary sense, but are rather ecotones—transition zones between aquatic and terrestrial systems.

Turloughs occur from Donegal to Limerick, extending in a broad curve through Sligo, Leitrim, Roscommon, Longford, Mayo, Galway, Clare, Tipperary and Kilkenny (Coxon 1987a, b; J.N.R. Grainger, pers. comm.). Coxon (1987a) has identified some 60 extant turloughs of at least 10ha, but many existing turloughs are small. Extensive turloughs existed in the past, now almost all are drained (Coxon 1987a; Baldock 1984).

While turloughs are a characteristic feature of karstic limestone country in the wetter parts of Clare, the Aran Islands, East Galway and Mayo, where they lie on pure Upper Carboniferous limestone, further east they may intergrade with temporary field ponds on well drained, shallow glacial drift, which have no well defined connection with ground water and which fill chiefly by rainfall and in winter. Rainfall conditions leading to filling midland turloughs occur mainly in the second half of the year (Grainger 1991). Indeed, in parts of Ireland they are called 'winter lakes'. Williams (1970) has described several types of turlough, including lowland types, which fill chiefly in winter and may not dry completely (e.g. the East Burren lowlands), and upland turloughs, which fill after heavy rain at any time of year (e.g. Lough Aleenaun in the Burren hills). There are also a number of permanent lakes with some turlough features (e.g. Lough Gealáin (grid ref. R 32 95) and Lough Bunny (R 37 97), Co. Clare; White Lake (N 51 72), Co. Westmeath).

Although turloughs are characteristic of Ireland and have been recognised as peculiar to it for 300 years (Nelson 1991), features analogous to turloughs are occasionally found elsewhere if the rainfall pattern, rock type and water table are suitable, and it is likely that more may be recognised. A turlough-like water body has recently been described at Pant-Y-Llyn in south-west Wales (C.A. Duigan, pers. comm.; Campbell *et al.* 1992), and another at Clot d'Espolla in Catalonia (Vila *et al.* 1989; Garcia-Gil *et al.* 1992; J.D. Reynolds, personal observation), although here the dry Mediterranean summer leads to a relatively predictable seasonal pattern of filling and emptying.

EXAMPLES OF STUDIED IRISH TURLOUGHS

Mullagh More

This turlough complex includes the extensive Knockaunroe turlough (30ha) and a number of smaller turloughs, a fen-turlough, a permanent lake (Lough Gealáin, 19ha) with a deep central doline and extensive marginal shallows that dry out (Byrne and Reynolds 1982), and a small permanent pond (Byrne 1981; Reynolds 1979; 1982). These oligotrophic waters (Byrne *et al.* 1989) are hydrologically linked to the Lower Fergus River (Drew 1990) and have aspects in common with some of the lakes and turloughs of the Gort lowlands, from Garrylands to Lough George. *Potentilla fruticosa* was first noted in a turlough situation from here (Lucas, letter of 1740, in Nelson 1991), while the characteristic turlough moss *Cinclidotus fontinaloides* occurs as an epiphyte in a small, heavily vegetated turlough (Coker 1993).

Carran

The extensive turlough of Carran (150ha; 3km long) lies in a polje (Corbel 1952; 1957). It is underlain by low-permeability cherty limestone and glacial deposits, and fills irregularly by marginal springs, retaining some water for 60–75% of the year (Drew 1990) and draining underground to the Fergus River. However, the Carran depression shows few of the features characteristic of turloughs, such as *Cinclidotus* on rocks, or zonation of vegetation (D'Arcy 1992).

Rahasane

This extensive turlough (275ha), representative of the larger East Galway turloughs of the past, is chiefly fed from the overflowing Dunkellin River, from which it can fill within two days, although it drains much more slowly. There are dense central stands of reeds and *Chara* spp, and *Cinclidotus* is widespread on boulders. The heavily grazed sward is chiefly composed of flood-tolerant species, e.g. *Agrostis stolonifera, Poa trivialis, Potentilla anserina* (Sheehy Skeffington 1985), and it is noted for waterfowl, including wigeon, Bewick and whooper swans (An Foras Forbartha 1981), while the small overflow arm was the first Irish site for the shrimp *Tanymastix stagnalis* (Young 1975; 1976).

TURLOUGH HABITATS AND THEIR BIOTA

While many inhabitants of turloughs are widespread elsewhere (Table 1), the biota of turloughs is well adapted to the special nature of their environment, with strategies including aerial adult forms, production of resting stages, resistance to dessication or an amphibious life-style. Organisms adopt different habitats in the turlough depending on their sensitivities to depth, temperature, or frequency and pattern of flooding.

Swallow-hole

A defined sink, swallow hole or '*slugaire*' region is present in some turloughs where the rock is near the surface. It is usually rocky and often permanently wet, frequently contains aquatic macrophytes such as *Nasturtium* and *Callitriche* species, and is a refuge for many organisms. The rocks provide shade and attachment surfaces (e.g. for the moss *Fontinalis antipyretica*). The water present may be cool, at or close to ground-water temperature (8°–10°C), and there may be cryptobionts or cave dwellers, sometimes washed out when the flow is strong. Permanent dwellers at the ground-water/surface-water interface may show reduced seasonality, like those in isothermal springs (Fahy 1975). Sink-holes may also hold small populations of larger, longer-lived aquatic organisms, e.g. *Gasterosteus aculeatus, Gammarus duebeni* or *Polycelis nigra* (Reynolds 1982), than are generally typical of turloughs.

Floor

The turlough may be floored with mud, marl, sand or vegetation, either aquatic or terrestrial, depending on factors such as water and substrate chemistry, depth and frequency or duration of filling. Burrowing invertebrates (e.g. *Pisidium*) may survive dry periods in the substrate.

Table 1—Some taxa characteristic of turloughs in Ireland, but not necessarily restricted to these ecotones.

Group	*Taxon*	*Reference*
ALGAE	*Oedogonium* sp.	Scannell 1972; Reynolds 1983
MUSCI	*Cinclidotus fontinaloides*	Praeger 1932
ANGIOSPERMS	*Viola persicifolia*	Praeger 1932
	Potentilla fruticosa	Praeger 1932
	Potentilla anserina	Coxon 1987b
TURBELLARIA	*Polycelis nigra*	Reynolds 1985
MOLLUSCA	*Lymnaea palustris*	Byrne 1981
INSECTA	*Cloeon simile*	Byrne 1981
CRUSTACEA	*Tanymastix stagnalis*	Young 1975; 1976; Grainger 1976; 1991
	Eurycercus glacialis	Duigan and Frey 1987a; 1987b
	Pleuroxus trigonellus	Duigan 1992
	Pleuroxus laevis	Duigan 1992
	Diaptomus castor	Grainger 1966; Ali *et al.* 1987
	Diaptomus cyaneus	Grainger 1976; Ali *et al.* 1987
	Diaptomus wierzejskii	Grainger 1976; Ali *et al.* 1987
	Diaptomus gracilis	J.N.R. Grainger, pers. comm.
	Cypris puber	J.N.R. Grainger, pers. comm.
	Gammarus duebeni	Reynolds 1982; 1985

The vegetation of turloughs has been analysed by Coxon (1987b), who shows a division into groups dominated respectively by wetland, calcareous grassland or ruderal plant species, reflecting the morphology, soils and pattern of flooding. Many western turloughs show distinctive zonation of herbaceous perennials (Praeger 1932; 1950; Ivimey-Cook and Proctor 1966; O'Connell *et al.* 1980; Webb and Scannell 1983), with aquatic or semi-aquatic plants (e.g. *Ranunculus* and *Potamogeton* spp, *Polygonum amphibium, Mentha aquatica*) extending away from the sink-holes and gradually giving place to a sward, where *Potentilla anserina* and sedges such as *Carex panicea* are typical and several *Viola* spp may be found. Shrubs such as *Frangula alnus* and *Potentilla fruticosa* may occur closer to the upper limits. On rocks may be seen *Cinclidotus fontinaloides,* the characteristic black turlough moss. In some western turloughs there is a transition to fen-type vegetation, including sedges and *Menyanthes trifoliata* (Reynolds 1982; 1985a; Young 1976). A spell of warm weather and a low and gradually decreasing water level will encourage the growth of filamentous algae, forming felts that bleach white on drying and are known as algal paper (Scannell 1972; Reynolds 1983).

While the aquatic faunal community has not received similar detailed treatment, a number of individual groups have been surveyed, such as entomostracans, beetles and molluscs (Ali *et al.* 1987; Bilton 1988; Byrne 1981;

Byrne *et al.* 1989; Duigan 1988; 1992; Grainger 1966; 1976; Lansbury 1965; Reynolds 1979; 1982; 1985b). As many turloughs are oligotrophic, receiving little allochthonous input apart from periodically flooded terrestrial macrophytes (Byrne *et al.* 1989), and as the initiation, duration and extent of flooding is unpredictable, the aquatic fauna may be sparse, only proliferating in favourable periods of warmth and with some water present.

When the turlough is full, the sward becomes a benthic habitat for gastropods (Donaldson *et al.* 1979), mayfly nymphs and chironomid larvae, usually restricted in diversity and rather sparsely distributed. Oligotrophic turloughs are relatively unfavourable habitats for even the euryoecic gastropods, and their growth rates are among the slowest known (Byrne *et al.* 1989).

The unpredictability of the aquatic fauna in time and space makes an inventory of the more interesting sites difficult. Much of the perceived value of turloughs comes from their importance as wintering or feeding sites for various wild duck, swans and waders (Ruttledge and Ogilvie 1979; Scott 1980) but these, like most aquatic invertebrates, may be neither vital nor predictable components of the turlough ecotone.

Open water

The open water overlying the short sward is the most extensive aquatic habitat. This is often oligotrophic and may be sparsely populated, but some spring-filling turloughs may show a bloom of planktonic algae and associated cladocerans, perhaps related to fertiliser addition to the surrounding sward (J.N.R. Grainger, pers. comm.). Outside the winter, turloughs may contain rich and characteristic entomostracan communities (Table 1), chiefly chydorids (Duigan 1988; 1992; Reynolds 1985b) and copepods (Ali *et al.* 1987; Grainger 1966; 1976). Many are coolwater species; most have resistant resting eggs, while some (e.g. *Tanymastix stagnalis, Eurycercus glacialis*) are obligate dwellers in temporary water bodies because their unprotected nature and large size would render them vulnerable to fish predation. Food limitation may restrict population size in *Tanymastix* (Grainger 1991).

TURLOUGH CONSERVATION

Threats and problems

The chief threats to turloughs are land drainage and pollution. Wetland drainage has been a feature of Ireland since the Famine, with up to 40,000 employed following the Drainage Act of 1842, resulting in the rapid drainage of 100,000ha (Baldock 1984; Bruton and Convery 1982). Drainage in the nineteenth century eliminated many of the great turloughs of East Galway in the creation of the Clare River as it is today. A second burst of activity followed the Arterial Drainage Act of 1945. The most recent projects have received European funding under FEOGA and the Regional Development Fund, and some environmental protection has also been invoked.

Recent schemes most affecting turloughs have been the Corrib–Clare drainage (1954–64), draining over 21,000ha of farmland, and the smaller Corrib–Headford and Corrib–Mask–Robe schemes (Baldock 1984; Bruton and Convery 1982). The hydrological complexity of these catchments is illustrated in Drew (1992). Seven sites of importance for Greenland white-fronted geese and Bewick and whooper swans were drastically affected (Scott 1980; Ruttledge and Ogilvie 1979), and drainage may also have curtailed the numbers of breeding black-necked grebe. The 40,000ha Dunkellin catchment has been under continued threat of drainage (Baldock 1984) and the river as it leaves Rahasane turlough was partly drained in 1992. Conversely, some field turloughs in the midlands have become permanently flooded, owing to compaction of the soil by the heavier breeds of cattle that have replaced sheep (J.N.R. Grainger, pers. comm.).

In regard to water quality, the complex hydrology, thin patchy soils and lack of filtration of water within the limestone aquifer make the Burren ground water particularly vulnerable to contamination (Drew 1990; 1992). Concern has been expressed about the possibility of sewage pollution of the Mullagh More turloughs from the proposed National Park Interpretative Centre. However, the situation is not merely local: parts of the Eastern Burren lie in the headwaters of the Fergus River, and silage effluent, septic tank overflow or even decaying animal carcasses anywhere in the catchment upstream of Mullagh More could have deleterious effects on the lakes and turloughs within the Burren National Park. While the severity would depend on the nature of the pollution, it should be remembered that much of the biota of turloughs is tolerant of changes in temperature and oxygen. As the karst of this area is generally fast-flushing, as evidenced by the rapid filling and level changes of the turloughs, the effects of a pollution event may be short.

Finally, although the animal community of turloughs is often sparse and unpredictable, species introductions could have a major effect. A three-year-old gravid perch was recently captured in Lough Gealáin, presumably released as a fry by a fisherman. If this had been allowed to mate and spawn, it could have had major effects on invertebrates and, through these, on phytoplankton and water clarity.

Conservation issues

As turloughs are almost peculiar to Ireland, and progressively disappearing as a result of drainage and other processes, conservation of at least the best representative examples should be a priority. However, there needs to be agreement as to what is to be conserved, the whole ecotone or a subset of the biota—vegetation, birds or invertebrates. The zonation of vegetation finds its fullest representation in only a few turloughs, and the majority are unimportant for waterfowl, perhaps because of size, depth and marginal vegetation. *Tanymastix stagnalis* occurs sporadically in a number of turloughs, of which Rahasane may be an atypical habitat (Grainger 1991). *Eurycercus glacialis* is similarly scattered (Duigan and Frey 1987a, b), while entomostracans may be most abundant in small pockets of water, and many typical turlough organisms are widely distributed outside these systems (Table 1).

While detailed regional studies of turloughs exist (e.g. Coxon 1987a; 1987b), a comprehensive scheme to conserve turloughs would first require an inventory to be made of all periodically inundated areas in each region, with an analysis of their catchment relationships, before categorising their most interesting features in order to select suitable examples of each type. However, human value systems do not treat all species equally, and certain birds, mammals and flowering plants would tend to endow a turlough with a higher perceived conservation importance than the majority of invertebrates. This problem permeates the entire question of site selection, acquisition and conservation perhaps even more than the required legal instruments. Finally, conservation of turloughs, being ecotones, requires sensitive management of the regional catchment, both surface and underground.

ACKNOWLEDGEMENTS

I am most grateful to my colleagues for their help and for freely sharing their information on turloughs, in particular to Dr Catherine Coxon, Dr David Drew, Dr Catherine Duigan, Professor Jack Grainger, Mrs Sylvia Reynolds and Dr John Rochford.

REFERENCES

ALI, T.H., HOLMES, J.M.C. and GRAINGER, J.N.R. 1987 *Diaptomus cyaneus* Gurney, a freshwater copepod new to Britain and Ireland. *Irish Naturalists' Journal* 22, 240–1.

AN FORAS FORBARTHA 1981 *Areas of scientific interest in Ireland.* Dublin. An Foras Forbartha.

BALDOCK, D. 1984 *Wetland drainage in Europe.* London. Institute for European Environmental Policy.

BILTON, D.T. 1988 A survey of aquatic Coleoptera in central Ireland and in the Burren. *Bulletin of the Irish Biogeographical Society* 11, 77–94.

BRUTON, R. and CONVERY, F.J. 1982 Land drainage policy in Ireland. *Policy Research Series* No. 4. Dublin. Economic and Social Research Institute.

BYRNE, R.A. 1981 Ecological comparisons of three water bodies in the Burren district of County Clare. MSc Thesis, Department of Zoology, University of Dublin.

BYRNE, R.A. and REYNOLDS, J.D. 1982 Preliminary notes on a doline in Lough Gealáin, the Burren, Co. Clare. *Irish Naturalists' Journal* 20, 375–7.

BYRNE, R.A., REYNOLDS, J.D. and McMAHON, R.F. 1989 Shell growth, reproduction and life cycles of *Lymnaea peregra* and *L. palustris* (Pulmonata: Basommatophora) in oligotrophic turloughs (temporary lakes) in Ireland. *Journal of Zoology, London* 217, 321–39.

CAMPBELL, S., GUNN, J. and HARDWICK, P. 1992 Pant-y-llyn—the first Welsh turlough? *Earth Science Conservation* 31, 3–7.
COKER, P.D. 1993 Epiphytic *Cinclidotus fontinaloides* in the Burren, Co. Clare, Ireland. *Journal of Bryology* 17, 509–11.
CORBEL, J. 1952 Un région karstique d'Irlande: Le Burren. *Revue de Géographie de Lyon* 27, 19–33.
CORBEL, J. 1957 Les karstes du nord-ouest de l'Europe et de quelques régions de comparaison. *Mémoires et Documents de l'Institut des Études Rhodamiennes de l'Université de Lyon* 12, 1–530.
COXON, C.E. 1987a The spatial distribution of turloughs. *Irish Geography* 20, 11–23.
COXON, C.E., 1987b An examination of the characteristics of turloughs, using multivariate statistical techniques. *Irish Geography* 20, 24–42.
D'ARCY, G. 1992 *The natural history of the Burren.* London. Immel Press.
DONALDSON, F., DONALDSON, F.L. and McMILLAN, N.F. 1979 A note on the Mollusca of three turloughs. *Irish Naturalists' Journal* 19, 400–1.
DREW, D.P. 1990 The hydrology of the Burren, Co. Clare. *Irish Geography* 23, 69–89.
DREW, D.P. 1992 Water resource management and water tracing in a lowland karst aquifer, western Ireland. In H. Hotzl and A. Werner (eds), *Tracer hydrology,* 221–5. Rotterdam. A.A. Balkema.
DUIGAN, C.A. 1988 The Cladocera of Lough Ree and neighbouring waterbodies. *Bulletin of the Irish Biogeographical Society* 11, 100–13.
DUIGAN, C.A. 1992 The ecology and distribution of the littoral freshwater Chydoridae (Branchiopoda, Anomopoda) of Ireland, with taxonomic comments on some species. *Hydrobiologia* 241, 1–70.
DUIGAN, C.A. and FREY, D.G. 1987a *Eurycercus glacialis,* a chydorid cladoceran new to Ireland. *Irish Naturalists' Journal* 22, 180–3.
DUIGAN, C.A. and FREY, D.G. 1987b *Eurycercus glacialis* in Ireland (Cladocera, Chydoridae). *Internationale Revue der gesamten Hydrobiologie* 72, 235–49.
FAHY, E. 1975 The biology of a thermal spring at Enfield, Co. Meath, with some observations on other Irish thermal springs. *Proceedings of the Royal Irish Academy* 75B, 111–23.
GARCIA-GIL, L.J., BORREGO, C.M., VILA, X., BAÑERAS, L. and COLOMER, J. 1992 Visit to the aquatic ecosystems of Lake Banyoles karstic area. In J. Catalan and J.L.I. Pretus, *Mid-Congress Excursions XXV SIL Congress* 7, 1–13. Barcelona. University of Barcelona.
GRAINGER, J.N.R. 1966 *Diaptomus castor* in County Meath. *Irish Naturalists' Journal* 15, 211.
GRAINGER, J.N.R. 1976 Further records for the fairy shrimp *Tanymastix stagnalis* (L.). *Irish Naturalists' Journal* 18, 326.
GRAINGER, J.N.R. 1991 The biology of *Tanymastix stagnalis* (L.) and its survival in large and small temporary water bodies in Ireland. *Hydrobiologia* 212, 77–82.
IVIMEY-COOK, R.B. and PROCTOR, M.C.F. 1966 Plant communities

in the Burren, Co. Clare. *Proceedings of the Royal Irish Academy* 64B, 201–311.

LANSBURY, I. 1965 Notes on the Hemiptera, Coleoptera, Diptera and other invertebrates of the Burren, Co. Clare and Inishmore, Aran Islands. *Proceedings of the Royal Irish Academy* 64B, 89–115.

NELSON, E.C. 1991 *The Burren.* Anglesey. Boethius Press.

O'CONNELL, M., RYAN, J.B. and MacGOWRAN, B.A. 1980 Wetland communities in Ireland: a phytosociological review. In P.D. Moore (ed.), *European mires.* London. Academic Press.

PRAEGER, R.L. 1932 The flora of the turloughs: a preliminary note. *Proceedings of the Royal Irish Academy* 41B, 37–45.

PRAEGER, R.L. 1950 *Natural history of Ireland.* London. Collins.

REYNOLDS, J.D. 1979 Swallow holes and turloughs. In F. O'Gorman (ed.), *The Irish wildlife book,* 58–60. Dublin. John Coughlan.

REYNOLDS, J.D. 1982 Ecology of turloughs (vanishing lakes) in the Burren, Western Ireland. *Transactions of the International Congress of Game Biologists* 14, 183–8.

REYNOLDS, J.D. 1983 'Algal paper' on Inishmore, Aran Islands. *Irish Naturalists' Journal* 21, 50.

REYNOLDS, J.D. 1985a Karstic freshwater habitats of the Aran Islands, Co. Galway, with preliminary notes on their fauna. *Irish Naturalists' Journal* 21, 430–5.

REYNOLDS, J.D. 1985b Cladocera from Burren turloughs at Mullagh More, Co. Clare. *Bulletin of the Irish Biogeographical Society* 9, 51–4.

RUTTLEDGE, R.F. and OGILVIE, M.A. 1979 The past and current status of the Greenland white-fronted goose in Ireland and Britain. *Irish Birds* 1, 293–363.

SCANNELL, M.J.P. 1972 'Algal paper' of *Oedogonium* sp., its occurrence in the Burren, Co. Clare. *Irish Naturalists' Journal* 17, 147–52.

SHEEHY SKEFFINGTON, M.J. 1985 Vegetation productivity in Rahasane turlough, Co. Galway. *Bulletin of the British Ecological Society,* 20–3.

SCOTT, D.A. 1980 *A preliminary investigation of wetlands of international importance for wildlife in West Europe and North West Africa.* Slimbridge. IWRB.

VILA, X., ABELLA, C.A. and BRUSI, D. 1989 Caracterització morfologica i limnologica del Clot d'Espolla i les surgencies del Pla d'Usall. *Scientia Gerundensis* 14, 23–42.

WEBB, D.A. and SCANNELL, M.J.P. 1983 *Flora of Connemara and the Burren.* Cambridge University Press.

WILLIAMS, P.W. 1970 Limestone morphology in Ireland. In N. Stephens and R.E. Glasoch (eds), *Geographical studies in honour of E. Estyn Evans,* 105–24. Belfast. Queen's University Press.

YOUNG, R. 1975 *Tanymastix stagnalis* (L.) new to Britain and Ireland. *Irish Naturalists' Journal* 18, 203.

YOUNG, R. 1976 *Tanymastix stagnalis* (L.) in Co. Galway, new to Britain and Ireland. *Proceedings of the Royal Irish Academy* 76B, 369–78.

In: J.D. Reynolds (ed.) 1996 *The conservation of aquatic systems*, 47–55. Dublin. Royal Irish Academy.

IRISH MARINE HABITATS

James G. Wilson[1] and Ian Lawler[2]

[1]Environmental Sciences Unit, Trinity College, Dublin
[2]Department of Zoology, Trinity College, Dublin

ABSTRACT

Ireland's varied marine habitats, near the centre of the boreal region at the northern limit for warm water lusitanian species and at the southern limit for arctic species, plus rare features like the limestone pavements on the west coast, offer great scope for rational conservation and management.

Survey results indicate that the most extensive habitat in the littoral zone is sand, followed in almost equal proportions by mud, rock and gravel. In the Irish Sea habitats tend to become more and more muddy from south to north. A sublittoral survey of Northern Ireland, which concentrated mainly on the rocky areas, has provided an unmatched database of habitats and species.

Habitats of international importance include the west coast maerl beds and limestone pavements with *Paracentrotus*. Only one marine locality, Lough Hyne, has been declared a National Nature Reserve. While there are Areas of Scientific Interest (ASIs) in the Republic of Ireland and Sites of Special Scientific Interest (SSSIs) in Northern Ireland, none has been declared in the sublittoral.

The recorded marine fauna of the Irish coast shows little difference in species composition from that recorded from the UK. However, there are records of species rarely reported elsewhere. In addition, the estuaries and mud-flats of the Irish coast provide overwintering grounds for nationally and internationally important numbers of wildfowl.

INTRODUCTION

Ireland is a comparatively small island on the edge of the European continental shelf. For its area, it has one of the longest coastlines in the world, yet the sea and the marine environment have never received their due level of interest.

In the nineteenth century and the early twentieth century, Irish biologists like Thomson in Belfast played an important part in enumerating and naming considerable numbers of marine species, and workers such as Holt (1909) regularly surveyed areas off the coast. There followed a considerable decline in activity, amounting to a cessation in some areas. Indeed for a while there was no national research vessel and it was not until the end of the 1980s that this need was reluctantly acknowledged.

In recent years, there has been a noticeable increase in interest in the marine environment, the Irish government signalling its intentions with the formation in 1987 of a new Department, Roinn na Mara, to take responsibility for this area, followed in 1992 by the creation of a Marine Institute. This has gone along with heightened public interest and participation in schemes such as Coastwatch and the Blue Flag campaign, which have received the backing of the EC, and in general it has been the implementation and enforcement of EC directives and policies that have provided much of the drive for the legislation in marine environmental matters.

ZOOGEOGRAPHY

Ireland lies in the boreal Atlantic region, the centre of which is usually considered to be the North Sea (Ekman 1953). However, the extension of the Gulf Stream up the Irish west coast extends the area of influence of the Mediterranean–Atlantic coastal waters around the north coast of Ireland, such that, in comparison, the fauna of the east coast in the Irish Sea lacks some of the colourful species found off the south and west.

The distribution of some of the large spectacular species, such as the crawfish *Palinurus vulgaris* and the bivalve *Pinna fragilis,* illustrates this effect. The latter is predominantly a southern species, yet it has been reported along the west coast of Ireland and Scotland as far north as Shetland. However, apart from one old record from Liverpool Bay, it was thought that live specimens did not occur in the Irish Sea (Seaward 1982), although one has just been found off Belfast (A. Davison, N.I. Aquarium, personal communication). Other less obvious species belonging to this group include the trumpet anenome *Aptaisia mutabilis,* the red mouth goby *Gobius cruentatus,* red dead men's fingers *Alcyonium glomeratum,* and the sea slug *Greilada elegans.*

The northern flow of the Gulf Stream also acts as a highly effective barrier to the southward penetration of arctic-boreal fauna. Records such as those for the squid *Sagittatus (= Todarode) sagittatus* in the Irish Sea—and off the Irish coast in general—are to be considered at best seasonal and at the southern end of the range (Seaward 1982). Another species illustrating this is the northern sea fan *Swiftia pallida,* which is known from one Irish record in the Kenmare River, all other sites from which it is known being from western Scotland. The burrowing anenome *Arachnanthus sarsi,* the stone crab *Lithodes maja* and the purple sun star *Solaster endeca* are further examples of this group. These examples illustrate the strategic position of Ireland at the conjunction of the zoogeographical provinces, and emphasise the potential for marine conservation.

HABITATS, COMMUNITIES AND SPECIES

Ireland, owing to its position and underlying geology, possesses a wide variety of marine habitats from soft intertidal muds to deep rocky cliffs and caves. In the littoral zone, information has been sent in by Coastwatch volunteers over the past four years, giving details on, *inter alia,* habitat type (see Table 1). Sedimentary environments—in particular sand and mud—account for the major proportion of habitats sampled in the surveys, but it must be borne in mind that the surveys themselves do not represent an unbiased sample of the Irish coastline. The concentration of survey participants on the east coast, along which the predominant shore type is one of long sandy bays interspersed with rocky headlands, would tend to magnify the relative proportion of sandy habitats. On the west coast, rocky shores predominated in certain areas, but the difficulty of access to some of these areas may have biased the surveys in favour of the more accessible beaches.

Table 1—Habitat types in Ireland's littoral zone (data from COAST-WATCH surveys 1989–92).

Survey year	*Rock*	*Boulders*	*Gravel*	*Sand*	*Mud*	*Other*
1989	28	19	29	41	27	4
1990	21	12	18	26	20	2
1991	26	13	28	44	30	2
1992	24	19	24	48	25	2
Average	24.8	15.8	24.8	39.8	25.5	2.5

In the sublittoral, the northern part of the Irish Sea has been particularly well studied, the southern part less so, but enough information exists to calculate relative proportions and areas of each habitat type (Table 2). There is a gradual fining of the sediments from south to north, with the southern part in particular dominated by gravel and gravel mixtures. Large patches of mud are found only in the deeper areas and off Dundrum Bay in the northern part of the Irish Sea. The predominance of gravel reflects the strong tidal currents throughout most of the Irish Sea.

As part of the Irish Sea Study Group report (ISSG 1990), the available literature on species distribution was drawn together along with some recent investigations (Mackie 1990). However, on the Irish side, the majority of the work has concentrated in and around Dublin Bay (e.g.Wilson 1983; Wilson and Shelley 1986), and in particular on the sludge dumping site (Walker and Rees 1980).

Comparable data for the south and west coasts of Ireland are less plentiful, at least in terms of the overall picture, although attention has focused from time to time on specific locations. Surprisingly little has been published on

Table 2—Irish Sea habitats (data from ISSG Report 1990).

Habitat type	*% Cover*	*Area (km^2)*
Mud	3.6	1,836
Muddy sand	7.4	3,774
Sand	15.3	7,803
Gravel/sand/mud	42.5	21,675
Gravel	31.1	1,5861

the littoral communities, with the exception of places such as Bantry Bay (Crapp 1973), which subsequently received considerable attention in the wake of an oil spill (Myers *et al.* 1980).

In the sublittoral some selected areas, such as Killary Harbour (Keegan and Mercer 1986), Kilkieran Bay (Könnecker and Keegan 1983), Kinsale Harbour (Dineen *et al.* 1986; Keegan and King (this volume)), have been studied and the habitats mapped, but Galway Bay is one of the few locations for which both a comprehensive list of species and details of their biology are available (e.g. Bowmer 1982; Doyle 1988). Outside these areas, data are scanty and limited either to specific locations (as listed above), or to a restricted range of species, such as molluscs (Nunn 1990), or to both, e.g. marine molluscs of the Burren (Killeen and Light 1989). A notable exception would have to be made for Lough Hyne in south-west Cork, which has been investigated since the 1920s (Kitching 1967). A symposium in 1990 brought together many of the more recent investigations (Myers *et al.* 1991).

It is interesting to note that rocky habitats have received much more attention on the west coast, e.g. Costello *et al.*(1986), than on the east, reflecting the relative paucity of true rocky areas, as opposed to gravelly areas of high current speed, on the east coast.

An overall picture of habitats round the Irish coast would be possible only in the broadest terms, drawing on localised data like those mentioned above and on other sources of geological or sedimentological information,

Table 3—Recent new records of species in Irish waters, showing phylum, species and date recorded.

Phylum	*Species*	*Date recorded*
Coelenterata	*Pachycerianthus multiplicatus*	1977
Coelenterata	*Arachnanthus sarsi*	1984
Coelenterata	*Halcampoides purpurea*	1977
Coelenterata	*Caryophyllia inornata*	1989
Coelenterata	*Swiftia pallida*	1981
Coelenterata	*Aiptasia mutabilis*	1981
Porifera	*Homaxinella subdola*	1991
Echinodermata	*Antedon petasus*	1984

e.g. Geoghan *et al.* (1989). Such comprehensive listings of species distribution covering wide geographical areas as are available, like the atlases of algae of the British Isles (Norton 1985) and of the molluscs of the British Isles (Seaward 1982), reflect as much the lack of recording round the Irish coast as the true picture of distribution outside those few well known areas. Table 3 lists some of the recent records of species new to Ireland, by no means all of which are small, insignificant or easily overlooked.

The ISSG Report identified 24 distinct littoral habitats and 27 distinct near-shore sublittoral habitats in the Irish Sea. These can be grouped into broader categories along the lines of the bottom types listed above. To these can be added other habitats described by comissioned surveys of both the sublittoral and littoral of Northern Ireland (Erwin et al. 1990; Fuller et al. 1987 respectively) and yet more habitats found in the south and west. Tables 4 and 5 attempt to bring these data together.

Table 4—Simplified list of littoral habitats around Ireland.

Habitat type	*Characteristic species*	*Location/example*
Exposed bedrock and boulders	*Verrucaria*, barnacles, *Alaria*	Headlands, south-west coast
Moderately exposed bedrock and boulders	*Fucus* spp, *Patella, Littorina, Laminaria digitata*	Headlands, east coast
Sheltered bedrock and boulders	*Ascophyllum, Laminaria saccharina*	Sheltered embayments, harbours
Limestone pavements	*Paracentrotus*	Burren coast
Caves	Encrusting sponges	Rathlin Island
Cobble	None	Small bays, south and west coasts
Shingle	Amphipods in weed at strand line	South and West coasts; exposed
Coarse sand	Amphipods, *Nephthys*	Fanore strand, east coast south of Dublin
Sand	*Tellina, Cerastoderma*	Embayments, east coast. e.g. Dublin Bay
Mud	Scrobicularia, Macoma	Sheltered inlets
Estuaries	*Nereis diversicolor*	River mouths

The surveys of the littoral and sublittoral in Northern Ireland were carried out with a view to establishing both the range and the importance of the various habitats. From the data available for Ireland as a whole, it is possible to attempt some listing of the types of habitats (see Tables 4 and 5), but rather more difficult to establish their importance, either on a local, national or international scale. One field in which a quantitative scale of

Table 5—Simplified list of sublittoral habitats around Ireland.

Habitat type	*Characteristic species*	*Example*
Bedrock and boulders, strong current	*Tubularia, Sagartia, Halichondria*	The Nose, Lambay
Bedrock and boulders, lesser current	*Alcyonium, Metridium, Antedon*	Rockabill
Bedrock and boulders, slight current/waves	*Laminaria hyperborea,* papillate sponges, *Eunicella*	Mullaghmore, Co. Sligo
Cliffs and overhangs	*Dercitus, Parerythropodium*	L. Hyne
Cobble, shell gravel, mixed	*Ophiocomina, Ophiothrix* beds	Dalkey Sound
Clean coarse sand	*Ammodytes*	Kish Bank
Medium sand	*Echinocardium, Corystes*	North Lambay
Maerl beds	*Aureliana, Mesacmea, Neopentadactyla*	Kilkieran Bay
Muddy sand	*Virgularia, Cerianthus*	Dublin Bay
Mud	*Amphiura, Nephrops*	Galway Bay, Lambay Deep

conservation importance exists and is used is bird conservation. A certain number of birds, the number varying according to the species, qualifies that location to be judged of national or international importance. Table 6 lists some examples of estuarine birds that attain those criteria at Bull Island, Dublin Bay.

Table 6 illustrates clearly that it is possible for even a small restricted area such as Bull Island, which is also under considerable human pressure, to be of international conservation importance. It also shows that it is possible to draw up such a scale for decision making in which relative numbers (compare e.g. dunlin and brent goose) can be taken into account. It is perhaps no surprise that of the marine or coastal ASIs in Ireland, almost three-quarters have been so designated on account of the bird life. (Note that this is being reviewed at the time of writing.)

What is needed for the rest of the marine environment is a similar listing and weighting, which would take into account local, national and international importance for marine species. A local dimension is important, as it is from local usage that most of the pressures on the habitat are likely to originate, and control measures, however well intentioned are unlikely to succeed without local backing. Some organisations, for example those based around Skerries and Killiney, Co. Dublin, have taken responsibility for their local marine environment, and one of the reasons for the success of Lough Hyne as a Marine Reserve has been the support it has received locally.

Table 6—Selected bird species of national or international importance at Bull Island Dulbin Bay (data from Merne 1985).

Species	*Numbers*	*National*	*International*
Brent goose	1,500	200	200
Shelduck	800	750	1250
Widgeon	2,600	2,000	5,000
Shoveller	300	90	1,000
Dunlin	7,900	5,500	20,000
Knot	6,700	2,500	3,500
Curlew	1,900	1,000	3,000
Redshank	2,400	1,000	2,000
Oystercatcher	3,800	3,000	7,500
Grey plover	150	100	800

Coastwatch surveys from 1989 to 1992 report a generally increasing perception of littoral areas being under threat from erosion, beach mining, sewage pollution and dumping, and recreational abuse. Agriculture, aquaculture and industry were seen as minor threats. In the sublittoral considerable habitat damage is being caused in intensively fished areas.

On a national and international basis, much has still to be learned of the rarity of the majority of marine species, with exceptions being made in the case of mammals and birds. However, size alone is no guarantee of knowledge, as is evident from the lack of information surrounding species such as the basking shark in Irish waters. A start is being made in this regard through the BIOMAR project, from which a much more detailed knowledge of the distribution of marine species around Ireland will emerge. This sort of information should be used as the basis for a national classification of marine habitats, to be used to set priorities for conservation action.

REFERENCES

BOWMER, C.T. 1982 Aspects of the biology of *Amphiura filiformis* (O.F.Müller) (Echinodermata: Ophuiroidea). Unpublished Ph.D. Thesis, University College Galway.

COSTELLO, J., KEEGAN, B.F. and KÖNNECKER, G.F. 1986 Rocky subtidal assemblages on the west coast of Ireland. *Hydrobiologia* 142, 97–111.

CRAPP, G. B. 1973 The distribution and abundance of animals and plants on the rocky shores of Bantry Bay. *Irish Fisheries Investigation Series* B. no. 9, 1–35.

DINEEN, P., KEEGAN, B.F., CROWE, W.A., O'CONNOR, B., McGRATH, D., COSTELLO, M. and KÖNNECKER, G. 1986 Littoral and benthic investigations on the south coast of Ireland—1. Macrofaunal distribution patterns and their relationship to the sedimentary environment in Kinsale harbour. *Proceedings of the Royal Irish Academy* 86B, 81–106.

DOYLE, M. 1988 An ecological study of fish larvae in the plankton of coastal waters, west of Ireland. Unpublished Ph.D. Thesis, University College Galway (2 vols).

EKMAN, S. 1953 *Zoogeography of the sea*. London. Sidgewick and Jackson.

ERWIN, D.G., PICTON, B.E., CONNOR, D.W., HOWSON, C.M., GILLEECE, P. and BOGUES, M.J. 1990 *Inshore marine life of Northern Ireland. Survey report to Conservation Branch DoE(NI)*. Belfast. HMSO.

FULLER, I., TELFER, T. and WILKINSON, M. 1987 The Northern Ireland littoral survey. *Porcupine Newsletter* 3(10), 268–76.

GEOGHAN, M., GARDINER, P.R.R. and KEARY, R. 1989 Possibilities for commercial mineral deposits in the Irish offshore area. *Marine Mining* 8(3), 267–82.

HOLT, E.W.L. 1909 Report of a survey of the trawling grounds on the coasts of counties Down, Louth, Meath and Dublin—part 1. Record of fishing operations. *Scientific Investigations Fisheries, Ireland* 1911.

ISSG 1990 *Irish Sea Study Group reports* (4 vols). Liverpool University Press.

KEEGAN, B.F. and MERCER, J.P. 1986 An oceanographical survey of Killary harbour on the west coast of Ireland. *Proceedings of the Royal Irish Academy* 86B, 1–70.

KILLEEN, I. and LIGHT, J. 1989 Marine recording on the Burren. *Conchological Newsletter* 111, 242–5.

KITCHING, J. 1967 Ecological studies at Lough Hyne. *Advances in Ecological Research* 4,198–291.

KÖNNECKER, G.F. and KEEGAN, B.F. 1983 Littoral and benthic investigations on the west coast of Ireland—XVII. The epibenthic animal associations of Kilkieran Bay. *Proceedings of the Royal Irish Academy* 83B, 309–24.

MACKIE, A.S.Y. 1990 Offshore benthic communities of the Irish Sea. In D. Shaw (ed.), *ISSG Report, part 1: Nature conservation*, 169–218. Liverpool University Press.

MERNE, O. 1985 Macrofauna and their availability as a food resource for birds in the Shannon and Fergus estuaries. Unpublished M.Sc. Thesis, University of Dublin.

MYERS, A.A., LITTLE, C., COSTELLO, M.J. and PARTRIDGE, J.C. 1991 *The ecology of Lough Ine*. Dublin. Royal Irish Academy.

MYERS, A.A. SOUTHGATE, T. and CROSS, T.F. 1980 Distinguishing the effects of oil pollution from natural cyclic phenomena on the biota of Bantry Bay, Ireland. *Marine Pollution Bulletin* 11, 204–7.

NORTON, T.A. (ed.) 1985 *Provisional atlas of the marine algae of Britain and Ireland*. Huntingdon. ITE Biological Records Centre.

NUNN, J. 1990 Recording mollusca in Ireland: sea area 34. *Conchological Newsletter* 115, 330–3.

SEAWARD, T. 1982 *Conchological atlas of the British Isles*. London. Conchological Society.

WALKER, A.J.M. and REES, E.I.S. 1980 Benthic ecology of Dublin Bay in relation to sludge dumping: fauna. *Irish Fisheries Investigation Series B*, no. 22, 1–59.

WILSON, J.G. 1983 The littoral fauna of Dublin Bay. *Irish Fisheries Investigation Series B*, no. 26, 1–19.

WILSON, J.G. and SHELLEY, C. 1986 The distribution of *Nucula turgida* (Bivalvia: Protobranchia) from Dublin Bay, Ireland, and the effect of sediment organic content. *Journal of the Marine Biological Association of the United Kingdom* 66, 119–30.

In: J.D. Reynolds (ed.) 1996 *The conservation of aquatic systems*, 56–67. Dublin. Royal Irish Academy.

STRANGFORD LOUGH—CONSERVATION CONSTRAINED

P.J.S. Boaden

**School of Biology and Biochemistry,
Queen's University of Belfast Marine Biology Station,
Portaferry, Co. Down**

ABSTRACT

The conservation importance of Strangford Lough stems from its habitat- and species-diversity. About 80% of Irish Sea coastal and inshore habitats and 75% of Northern Ireland recorded marine species are represented in the Lough. Species abundances also merit international and national recognition. Movement towards effective conservation was triggered thirty years ago by responsible wildfowlers concerned for their sport. This led to the National Trust's Strangford Lough Wildlife Scheme, which widened toward overall management—the Trust *in loco parentis* through ownership and lease of foreshore rights being able to exert some influence on planning. At times the Scheme has contained considerable tensions, particularly in relation to wildfowling and refuge areas. Fisheries and recreation present growing problems requiring effective control.

Statutory designations include one Area of Outstanding Natural Beauty, seven National Nature Reserves and three Areas of Special Scientific Interest. The last led to controversy over public and private rights. Existing and proposed measures form 'nested' designations over most of the foreshore and natural fringing habitat. Two other major initiatives are underway, namely statutory designation as a Marine Nature Reserve and integration of wider interests through a Strangford Lough Management Committee.

The natural value of Strangford Lough constrains us to conserve it, but to date conservation has been constrained by conflict, compromise and lack of conducive legislation.

INTRODUCTION

Setting

Strangford Lough, Northern Ireland, is one of the largest sea loughs in the British Isles. It is almost entirely landlocked, being bounded on the east side by the Ards Peninsula and on the north and west by the bulk of County Down. Connection to the Irish Sea is through the Strangford Narrows, which is about 8km long and mostly less than 1km wide. There is a fairly narrow entrance sill, but the deepest part of the Narrows (66m) is at its northern end. The body of the Lough is 24km long, roughly on a north–south axis, and approximately 8km wide. Much of the Lough is less than 20m deep, but there is a Y-shaped central channel, mostly of 30–50m. About one-third of the Lough's 150km^2 total area is intertidal.

Spring tides range up to a maximum of just over 4m height, but average range is about 2.8m. However, even on average tides, some 350 x 10^{6}m^3 of water enter and leave the Lough. Thus tidal currents and turbulence can be spectacular, with speeds of up to 18km h^{-1} creating a whirlpool—the Routen Wheel—near the narrowest point. Currents lessen rapidly toward the north of the Lough, but are still often of 1–1.5km h^{-1}. The tidal hydrology is a dominant ecological factor, since it winnows and sorts particulate matter, thus controlling the Lough's substrata and benthic communities.

The Lough is largely underlain by Silurian greywacke siltstone and shale, but its general scenery results from its quaternary overlay of glacially moulded debris. Thus the Lough may be regarded as a drowned drumlin field and is fjardic rather than fjordic. Although the Lough is relatively sheltered, its littoral and infralittoral areas are subject to wave action in addition to tidal effects. Since bad weather in the area is dominated by winds with a westerly component, the eastern shores have borne its brunt. Thus on the Ards side of the Lough the intertidal and island drumlins have been largely eroded to form low-lying reefs, known locally as pladdies, and the sediments tend to be sandy; conversely on the western side many islands persist and shores are rather muddy. Thus wave action contributes to the hydrodynamic generation of the Lough's diversity.

Surveys

Northern Ireland's marine life has been studied intermittently for centuries and has been used by mankind for millennia. Harris (1744) considers resource use conflict between herring fisheries and kelp, i.e. seaweed ash, production around the County Down coast. Belfast's Victorian naturalists made considerable contributions to marine taxonomy in which studies of Strangford Lough's fauna played an important role. Knowledge of this fauna was summarised by Williams (1954) but has since expanded considerably. Much of the impetus for further studies has derived directly and indirectly from the Queen's University Marine Biology Station in Portaferry, which was originally established under the honorary directorship of G.M. Williams and subsequently functioned under the directorship of the author, who also served on the Department of the Environment for Northern Ireland (DOE) Nature Reserve Committee (and its successors) and the National Trust's Strangford Lough Committee. Documentation of the Lough's marine wildlife now chiefly resides on computer file in the Ulster Museum. Most of the information is derived from the Museum's own work, from their DOE-commissioned sublittoral survey around Northern Ireland (Erwin *et al.* 1990) and from a similarly commissioned littoral study by the Herriot-Watt University, in addition to the DOE's own supra-littoral habitat survey. Further information emanates from surveys conducted in relation to the Strangford Lough Scheme (see below) and various other sources. An excellent popular account of the Lough's wildlife is given by Brown (1990).

Status

The reports of the Irish Sea Study Group (ISSG) included an account of current moves towards standardising the classification of marine habitats and communities around Great Britain (Holt *et al.* 1990). Baxter and Boaden (1990) re-interpreted the available Northern Ireland data in this context, and subsequently A.J. Davison and P.J.S. Boaden in an unpublished report to the DOE rated the importance of the communities, habitats and species of Strangford Lough on the basis of local, Northern Ireland, national (UK and/or Irish) and international significance, these categories approximately corresponding to the UK conservation review grades 4, 3, 2 and 1 respectively (see Ratcliffe 1977). This information is summarised in Table 1.

Table 1—A summary of the wildlife importance of Strangford Lough, Co. Down, N. Ireland.

	In Irish Sea	*In Strangford*	*Significance Local/NI*	*Significance Nat./Internat.*
Number of Community types on ISSG/NCC basis				
Marginal	9	6	4	2
Littoral	24	18	10	6
Sublittoral	27	23	9	7

Northern Ireland Survey recorded species

	Total NI	*In Strangford*	*Exclusively SL*
Littoral	945	76%	7%
Sublittoral	956	72%	28%

Regular winter waders and wildfowl

Species	*Ir./UK significance*	*Internat. significance*
35	25	3

Other important features

Terns: 30% of all Ireland population

Common seal: 450–600 individuals, largest Irish population

Algae and marine invertebrates: over 2000 spp recorded, including many rarities

Fish: over 60 spp recorded; nursery area for various species, e.g. tope

Semantics

If 'conservation' is to be defined as 'the wide use of resources', three things are assumed, namely (in reverse order) the stock, reserve or means of supplying some want; utilisation; and the exercise of sound judgement (Onions 1944). To 'constrain' is to force, compel or oblige or even, according to Dr Samuel Johnson, 'to produce in opposition to nature' *(ibid.)*. If we accept the ethics of conservation as outlined in the World Conservation Strategy (International Union for the Conservation of Nature 1980), the species diversity and natural resource utilisation of Strangford Lough imply a moral commitment. Thus we are constrained by the nature of the Lough to conserve it. How then have people sought to signal attention to the Lough and to warn off adversaries?

SIGNALS OF CONCERN

Harris (1744) drew attention to the decline in herring fisheries in County Down, including Strangford Lough, and cited fisheries malpractice, kelp burning and predation as possible causes. His remedial proposals included mesh-size regulation, better refuse disposal and both seasonal and spatial apportioning of activities—all of which might meet with present-day approval, although the suggestion of killing whales, seals and seafish would not. It is not known whether any of these proposals were put into practice. Some other aspects of the early history of Strangford can be found in Boaden (1984).

Strangford Lough Wildlife Scheme

Current conservation management of the Lough may be dated from the early 1960s when several local wildfowlers became concerned that indiscriminate shooting was destroying their own sport and approached government and other sources to seek a regulatory mechanism. This led to the establishment in 1966 by the National Trust (NI) of a wildlife scheme within which wildfowling was to be controlled through a permit, wardening and refuge system; the Trust undertook by deed of covenant to consult with the appropriate shooting organisation, now represented by the British Association for Shooting and Conservation (BASC) on matters relating to wildfowling on the Lough, having acquired enabling foreshore rights and agreements.

A Joint Council of Wildfowling Clubs was formed to focus Lough wildfowling interests and aid in management. A Scheme Committee was established with a membership (originally of 13, now 20) made up of wildfowlers, naturalists, landowners and others interested in activities around the Lough. A full-time warden was appointed and an honorary wardening system established. The warden worked and reported to the Trust via the Scheme Committee, his day to day activities, ostensibly at least, being under the direction of a Wildlife Panel (originally four strong), whose duties included recommendations for and review of a refuge area system. Regular monthly winter bird counts were organised by P. Mackie, one of the original Scheme and Panel members; these counts probably represent the longest continuous biological data set in Ireland. Although much of the early work of the Scheme was directly related to wildfowling, it was realised even before the Scheme's institution that a much wider remit to promote protection of all the Lough's wildlife would have to be taken.

The original refuge areas remained largely unchanged from 1966 until reassessment in 1983, when two were enlarged (one absorbing three smaller refuges) giving six areas totalling about 16.5km^2 of shore and sea surface. These expanded areas caused considerable protest among wildfowlers and were only agreed to by them for a three-year trial period, partly to see if the expansion would help to halt the severe decline in wigeon numbers. Following review in 1986 the Wildlife Panel recommended that the extended system be retained or, more correctly, reinstituted, since following extensive pressure from wildfowlers the Trust had returned the larger of the extended refuges to its original size after the 1985–6 season. Since it was only in refuge areas that wigeon numbers had been maintained and since there appeared to be a need for more effective conservation of other species, the Panel also recommended two further refuges. Discussion of these proposals led to considerable conflicts within the scheme and to deterioration of the relationship between the National Trust and BASC. It only proved possible to reach a compromise by retaining the original refuges (one in its enlarged state) and adding one further refuge on a to-be-reviewed basis.

It is possible that the 'administratively minded' might feel well pleased with such an outcome; however, the local wildfowling community remained aggrieved with the Trust, believing further restrictions on them unjustified, and conversely the ornithological and naturalist lobby felt thwarted (by BASC) in their aims to conserve bird stocks. Several suggestions were made to take the matter to arbitration but the covenanting parties eventually compromised in 1988 with the establishment of a 'steering group to address the need for a major review of the Lough's winter wildfowl and waders'. This led to the formation of a Review Panel managed via the International Waterfowl and Wetlands Research Bureau (Slimbridge), the appointment of an independent Review Coordinator (Dr Palle Uhd Jepsen from the Danish Ministry of the Environment Wildlife Administration), the commissioning of reports, a review workshop (see Moser 1990) in June 1990 and a final report and recommendations by the co-ordinator (Jepsen 1990). Both parties (i.e. the National Trust NI and BASC) had previously agreed to stand by the recommendations unless there were 'substantial and compelling reasons' not to do so.

The recommendations were themselves substantial, proposing for example the retention of the existing refuges, which, with the addition of two extensive new areas, were to act as 'core wildlife zones'; the restriction of northern foreshore wildfowling to flood tide periods only; and some seasonally restricted shooting within two of the core wildlife zones.

BASC, and subsequently the Trust, found substantial and compelling reasons, entered closed-door discussions and emerged with agreed refuge proposals, which in terms of total area were roughly comparable to the pre-review situation. Thus it appears that the Jepsen proposals, which might have seen an estimated 50% or more extension of wildfowl protection, had been jettisoned. However, the Wildlife Scheme continues to function and it has certainly had considerable success in maintenance and protection of the environment and in helping to stimulate public awareness of the Lough. The Trust, as for all its 'properties', has produced a management plan (1985), which remains one of the best (although semi-private) sources of information about the Lough. Some of the Jepsen proposals regarding better management procedure are being implemented and the Trust's opinions on matters such as slipway and marina development are variously sought and promulgated. The Trust continues to have good relations with sailing and other organisations and these, for example, help to limit disturbance on nesting islands. However, an increasing number of 'unattached' power-boaters, jet-skiers, horse-riders etc. are using the Lough and are not necessarily so amenable to voluntary restriction.

Statutory designations

The gestation period of the Wildlife Scheme also saw the production of the first major amenity and conservation legislation for Northern Ireland, thus enabling the establishment of Nature Reserves (NR) and Areas of Outstanding Natural Beauty (AONB). Under the terms of this Amenity Lands Act (NI) 1965, five NRs containing diverse intertidal areas, coastal calcareous grassland, marginal vegetation and a freshwater pondage area were declared in the Strangford Lough area (see Boaden 1984). One of these, the Dorn, included the first (1976) statutorily protected sublittoral site in the British Isles. An extensive area around the Lough was designated an AONB, as was adjacent coastline (Lecale) to the south. The designation requires the Planning Service to pay particular attention to conservation and amenity aspects. The Strangford AONB is currently being revised under the superseding Nature Conservation and Amenity Lands (NI) Order 1985, which is administered by the Department of the Environment NI (DOE). Under this Order the five NRs have acquired the additional sobriquet of *National,* and two more NNRs have been added, one an interglacial raised cobble beach with appropriate vegetation, and one comprising the northern muddy-sand foreshore of the Lough. There is also a Wildlife (NI) Order 1985, which can afford various degrees of protection to particular species. This Order, for example, currently protects many birds and marine mammals, thus helping to conserve several species important to the ecology of the Lough.

The Nature Conservation and Amenity Lands Order also enabled the establishment of Areas of Special Scientific Interest (ASSI), a designation intended to 'achieve its aims by co-operation' and whereby 'landowners retain ownership and occupancy of their land and are usually encouraged to maintain it as heretofore. Landowners and occupiers are provided with a list of operations and activities which have the potential to lessen the nature conservation value of the site. These are not to be undertaken without consulting the Department' (DOE 1991a).

The Department's Countryside and Wildlife Branch has an ASSI survey team, which undertook a review and survey of Strangford's foreshore and fringing habitats (e.g. saltmarsh and scrubland). This led to the recommendation to designate virtually the entire intertidal zone and remaining natural fringing habitat. This was done in three stages. The first two declarations (1988) passed with little comment, but the third (1989) (incidentally coinciding with a fairly wealthy area of the Loughside community) caused a furore of accusations of high-handedness, trespass and draconic suppression of public and private rights. However, in due course criticisms, whether justified or not, were met, assuaged or withdrawn and the DOE were able to proceed with the designation.

SUBSTANTIVE ISSUES

Suprastructure

It has for long been apparent to those concerned with conservation of the Lough and its environs that there is considerable interaction between different parts of the Lough, its waters, hinterland, wildlife and human uses. Whereas it is obvious to the general public that major changes such as a tidal barrage (Northern Ireland Economic Council 1981) would cause fundamental ecological and sociological changes, it is not so apparent that the nature of the Lough is under constant attrition from *seemingly* small changes, such as slipway construction, increased fishing activity and growing disturbance through increased leisure. It is also clear to the concerned that, unless there is due evaluation and care, seemingly public (and often free) resources, such as views or beach access, may become privately appropriated, for example by building across them, and it is then too late to complain. For such reasons it is necessary not only that some management mechanism for the waters and sublittoral areas of the Lough be found, but also that this should be integrated into an overall pro-active policy of wise use of resources.

A substantial effort towards this end was made in the mid-1970s, representation by interested individuals and organisations resulting in the establishment of a Strangford Lough Working Group, which presented an unpublished report (commonly called the Wolsey Report) to the Nature Reserves Committee (DOE) in 1979. It recommended the establishment of overall management through a two-tier system basically comprised of a committee representing those with statutory and legislative powers and an advisory committee representing Lough users. This structure was not called into being, since the report coincided with Mrs Thatcher expressing anathema for quasi-autonomous non-governmental organisations. However, there were subsequently several meetings of a government-sponsored talk-shop, which gradually faded into silence, and a Strangford Lough Advisory Co-ordinating Council (SLACC), mainly comprised of representatives of interested organisations, established itself to pursue the aim of integrated management. More recently the Balfour (1984) Report on the status and administration of conservation in Northern Ireland recommended overall management of Strangford Lough, as did the House of Commons Environment Committee (1991) after visiting the Province.

When the Northern Ireland Wildlife and Amenity legislation (see above) was being redrafted, it was suggested (among others by the author, with Strangford and similar situations in mind) that provision should be made, not only for effective marine nature reserves (the 1981 Great Britain legislation already appearing lamentable) under complete control by the DOE, but also for areas jointly administered by them and the Department of Agriculture for Northern Ireland (DANI), who are the statutory body for Northern Ireland fisheries. These ideas were approved by the then Northern Ireland Assembly's Environment Committee, but lost (apparently since provisions substantially different from the previous GB legislation could not be drafted) after the Assembly was prorogued.

As a result of the ASSI controversy, various local residents founded a group called the Strangford Lough Nature Conservation Association (SLNCA), one of its initial prime objectives being to oppose the conservation designation concerned. The group prided itself on being democratically formed from among the local community, unlike SLACC, which they held to be undemocratic through its appointee system and not to reflect local opinion properly. The SLNCA expanded rapidly and took a very high profile through a series of public meetings and media reports. It was initially regarded with considerable unease by many already concerned with conservation of the Lough. However, it also achieved overt and covert help and influential sympathy, for example from some members of the DOE's Advisory Council for Conservation and the Countryside, for another of its prime objectives, namely ensuring local participation in any management suprastructure for the Lough.

The establishment by Ards Borough Council of the Northern Ireland Aquarium, Portaferry, has also been of importance in raising public awareness of the Lough's wildlife.

Spartina

The Wildlife (NI) Order 1985 also provides for regulating the introduction and spread of non-native species. There has been a long standing worry over the spread of cord-grass *Spartina anglica* in the Lough and various attempts to control it have been made by the National Trust, Ards Borough Council and the DOE. The last major attempt in 1981–3 led to a major court case, Cuan Sea Fisheries vs DOE 1989/90, since the area of control was adjacent to the Lough's main oyster culture area. It was alleged that, following spraying and resultant death of the grass, large quantities of previously trapped silt were released leading to smothering and loss of oyster production. The case was eventually settled out of court with extensive costs being awarded to the plaintiff. Subsequently, *Spartina* has respread in the area and further control action has been largely inhibited.

Shellfish and aquaculture

There has also been concern over possible detriment from the spread of aquaculture and, in particular, proposals to introduce trout and salmon cage farming to the Lough. However, following strong lobbying from naturalist interests, it was decided that fin-fish licences would not be granted. The current level of shellfish culture appears to be integrated in the local economy but not necessarily (see *Spartina*) in the Lough's ecology. It is known that overturning of boulders by winkle pickers causes some detriment to shore fauna and flora in some areas and changes resulting from an attempt at seaweed harvesting have been documented (Boaden and Dring 1980). Some effort is underway to establish scallop *Pecten maximus* culture in the Lough and it may be of interest to note that trade in the sea urchin *Echinus esculentus* is prohibited by schedule of the Wildlife Order.

Scraping the bottom

Since the late 1960s it has been apparent to marine biologists that trawling and dredging in the Lough are destroying or altering some bottom communities. This was originally noted in relation to a *Pecten* bed near the north end of the Narrows where dredging, by more powerful boats than had traditionally used the area, displaced bottom boulders that had sheltered the intervening sandy gravel; this was then scoured out, the bed becoming stoney and largely unsuitable for the 'clams' (personal observation). Thus the Wolsey Report recommended some restriction of fishing activity in the Lough, including regulating boat size. Similar alteration of bottom communities by dredging has been documented elsewhere (Caddy 1973).

A trawl fishery for 'queenies', *Aquipecten opercularis,* developed in the Lough in the mid-to-late 1980s and sadly has remained more or less uncontrolled. It has caused extensive damage to the unique horse-mussel communities in the central area of the Lough (Brown 1989), a large proportion is now devoid of epibenthic species, including the hydroids on which young queenies can settle. However, at present there is still some fishery, facilitated presumably by immigration of queenies from remaining unfished areas. A video, clearly illustrating the damage, was sponsored by the National Trust and called for a more rational approach to fisheries/conservation issues in the Lough. Fisheries interests deemed the video to be biased, possibly because of its title, which has been re-used as this section's subheading. The effects have subsequently been confirmed by further investigations (Industrial Science Division 1990) sponsored by DANI and the DOE, who continue to debate the issue as further damage occurs. Influence brought to bear through these departments, then under a common minister of state, and the opinion of the Crown Estate Commissioners that non-traditional activities that alter sea-bed structure could be prohibited (personal communication) 'saw off' an attempt to introduce suction dredging to the Lough in the late 1980s.

SECURING THE FUTURE

The move towards establishing marine nature reserves (MNRs) in Great Britain has been hampered rather than helped by the relevant legislation, which makes no provision for transfer of existing powers, right and functions to management (see Gubbay 1989). This has meant extensive and prolonged negotiations, particularly in relation to fisheries. It is unfortunate that the Northern Ireland legislation mainly mirrors that of Great Britain. However, a Strangford Lough Marine Nature Reserve has now been promoted or promulgated by at least three successive ministers of state. To pursue this aim the DOE in 1989 commissioned a study 'designed to provide a basic outline of the information on all aspects of the Lough and its management on which we will then be able to build' (DOE Conservation Service letter to interested parties). The subsequent report to the DOE (A.J. Davison and P.J.S. Boaden, The management of Strangford Lough: a study of the marine environment, its conservation and its exploitation) completed and revised in 1990, was prepared as requested in the format used by the GB Nature Conservancy Council (NCC 1988) for the preparation of site management plans. The report specifically considered various management options for the Strangford Lough area.

Following digestion of the report, the DOE prepared a consultation paper (DOE 1991a) outlining the Lough's importance and suggesting mechanisms, issues and objectives relevant to proposed management. Some management proposals, in particular the creation of an extensive MNR comprising all the Lough and its Irish Sea approaches and a two-tiered management structure, were closely reminiscent of the 1979 Wolsey Report. Public debate and response to the paper were largely favourable, but seem to have convinced the DOE that any management structure for the Lough could not succeed unless non-statutory voices were given an unequivocal role in management. Correspondingly it was recommended (DOE 1991b) that a single-tier structure, the Strangford Lough Management Committee, be established. This will work with the DOE towards the establishment of the Marine Nature Reserve and its management plan, also advising how this may be extended to cover the wider area and range of uses of the Lough's environs. SLACC has now disbanded, believing its objectives have largely been achieved or are safe with the new Committee. SLNCA has been accepted as having a contribution to make within the Committee and the National Trust appears relieved to be able to relinquish at least some of its former role *in loco parentis* for the Lough to the Committee.

It is hoped that the MNR draft management plan will be ready for consultation by autumn 1993. The Committee has had several meetings and a liaison officer has been appointed. There are rumours of some strong differences of opinion in relation to particular interests, both within and without the Management Committee, but it is hoped that these prove insubstantial in view of the considerable tasks that lie ahead in safeguarding the future of wildlife and resource use in Strangford Lough. Present designations and possible further declarations, for example, as a Ramsar site or as an Environmentally Sensitive Area, together with the proposed MNR status provide an overlapping protective meshwork or safety net for the Lough's wildlife, but true conservation requires public participation in and benefit from the Lough's well-being.*

REFERENCES

BALFOUR, J. 1984 *A new look at the Northern Irish countryside.* A report produced for Chris Patten, MP, parliamentary under secretary, the Department of the Environment, Belfast. HMSO.

BAXTER, J. and BOADEN, P. 1990 Northern Ireland. In Irish Sea Study Group, *The Irish Sea: an environmental review. Part 1: Nature conservation,* 83–102 + 2 maps. Liverpool University Press.

BOADEN, P.J.S. 1984 Strangford Lough: a conservation history. In D.W. Jeffrey (ed.) *Nature conservation in Ireland: progress and problems,* 57–66. Dublin. Royal Irish Academy.

BOADEN, P.J.S. and DRING, M.T. 1980 A quantitative evaluation of the effect of *Ascophyllum* harvesting on the littoral ecosystem. *Helgolander Meeresuntersuchungen* 33, 700–10.

BROWN, R.A. 1989 Bottom trawling in Strangford Lough: problems and policies. In *Proceedings of the third North Sea seminar,* 117–27. Amsterdam. Workgroup Noordzee.

BROWN, R.A. 1990 *Strangford Lough: the wildlife of an Irish sea lough.* Belfast. The Institute of Irish Studies, Queen's University.

CADDY, J.F. 1973 Underwater observations on trawls and dredges and some effects of dredging on a scallop ground. *Journal of Fisheries Research Board Canada* 30, 173–80.

DOE 1991a *Strangford Lough: a consultation paper.* Belfast. Environment Service, Department of the Environment (NI).

DOE 1991b *Strangford Lough: management structure.* Belfast. Environment Services, Department of the Environment (NI).

DOE 1992 *An introduction to Areas of Special Scientific Interest.* Belfast. Countryside and Wildlife Branch, Department of the Environment (NI).

ERWIN, D.G., PICTON, B.E., CONNOR, D.W., HOWSON, C.M., GILLEECE, P. and BOGUES, M.J. 1990 *The inshore life of Northern Ireland.* Belfast. HMSO.

GUBBAY, S. 1989 *Coastal and sea use management: a review of approaches and techniques.* Ross-on-Wye. Marine Conservation Society.

HARRIS, W. 1744 *The antient and present state of County Down.* Dublin. Exshaw. (Reprinted 1977. Ballynahinch. Davison.)

HOLT, R., FISHER, E. and GRAHAM, C. 1990 Coastal resources of the Irish Sea. In Irish Sea Study Group, *The Irish Sea: an environmental review. Part 1 Nature conservation,* 5–38 + 3 maps. Liverpool University Press.

*Note added in press: The Lough was designated a Marine Nature Reserve in July 1995.

HOUSE OF COMMONS ENVIRONMENT COMMITTEE 1991 *Environmental protection: environment issues in Northern Ireland.* London. HMSO.

INTERNATIONAL UNION FOR THE CONSERVATION OF NATURE AND NATURAL RESOURCES 1980 *World conservation strategy.* Gland, Switzerland. IUCN.

JEPSEN, P.V. 1990 *A review of the Strangford Lough Wildlife Scheme: final report and recommendations.* Slimbridge. International Waterfowl and Wetlands Research Bureau.

INDUSTRIAL SCIENCE DIVISION 1990 *The impact of commercial trawling on the benthos of Strangford Lough (interim report).* Belfast. Department of Economic Development.

MOSER, M.E. (ed.) 1990 *Strangford Lough Wildlife Scheme: report of the review workshop.* Slimbridge. International Waterfowl and Wetlands Research Bureau.

NATURE CONSERVANCY COUNCIL 1988 *Site management plans for nature conservation — a working guide.* Peterborough. NCC.

NORTHERN IRELAND ECONOMIC COUNCIL 1981 *The Strangford Lough tidal barrage: a feasibility study.* Belfast. NIEC.

ONIONS, C.T. 1944 *The shorter Oxford English dictionary* (3rd edn). Oxford University Press.

RATCLIFFE, D.A. 1977 *A nature conservation review.* Cambridge University Press.

WILLIAMS, G. 1954 Fauna of Strangford Lough and neighbouring coasts. *Proceedings of the Royal Irish Academy* 56B, 29–133.

In: J.D. Reynolds (ed.) 1996 *The conservation of aquatic systems*, 68–80. Dublin. Royal Irish Academy.

THE FAUNA OF KINSALE HARBOUR AND APPROACHES, SOUTH COAST OF IRELAND

Brendan F. Keegan and Pauline King

Department of Zoology, The Martin Ryan Marine Science Institute, University College, Galway

ABSTRACT

Since 1978, the littoral and sublittoral communities of Kinsale Harbour and its approaches have been the foci of an environmental impact study related to the discharges from a local pharmaceutical company. As major features of this study, natural and introduced substrates have been investigated for pattern in species settlement and community development, the soft-bottom fauna and associated sediments are characterised on an annual basis, and monthly hydrographic sampling is carried out across the study area.

In terms of the littoral and benthic fauna of the area, there is nothing to suggest that such minor changes as have been recorded lie outside the system's inherent capacity for natural change. The growing database continues to reinforce the early discovery that dimensions of natural stress are imposed on part of the fauna by the often very high suspended load of the river Bandon, and by occasional turbation from river flow and gales with a southerly to easterly aspect. Severe climatic conditions can exaggerate these stresses and bring about significant change in the 'more visible' intertidal (e.g. the sloughing away of mussel patches to be replaced by incursive algae). Overall, the system clearly has a high measure of resilience.

BACKGROUND

In 1978, the pharmaceutical company, Eli Lilly S.A. (Irish Branch) commissioned the Benthos Research Group of the Zoology Department, University College, Galway, to initiate an environmental impact study in the part of the Kinsale Harbour area that was to receive effluent from the company's proposed Dunderrow plant. This study, which has been variously pursued since that time, primarily focuses on the macrobenthic fauna of the shore and sea-floor, with some consideration of the water column and of associated sedimentary habitats where appropriate. There has been a phased approach to the work: a characterising survey (1978–81), a baseline investigation (1979–82), and a monitoring study (1982–ongoing).

The characterising survey was designed to ascertain in good detail the nature and distributional extent of the littoral and benthic communities in the general vicinity of the proposed outfall (see Dinneen *et al.* 1986). The overall survey was planned so that sufficient information would be generated to permit the selection of sites that, collectively, would be representative of the area at large and would also be suited to further, quantitative investigation.

The baseline investigation began in March 1979 and can be conveniently described in terms of its two broad habitat component, littoral and sublittoral.

Littoral

This expanded on the previous work and incorporated all of the more accessible study sites from the Old Head of Kinsale to Kinsale town, and along the eastern flank of the harbour to Ferry Point. A number of locations were selected for ongoing investigation and these were subsequently visited on a regular basis to describe the population structure and dynamics of certain prominent species. A control site, outside the immediate impact area, was studied in parallel.

Sublittoral

Based on the characterising survey, one hard- and three soft-bottom sites were selected for on-going study. Frequent remote (from surface craft) and *in situ* (diver-controlled) sampling was effected to identify processes and events in species settlement and community development, and to relate these to aspects of the physical/chemical environment. Emphasis was placed on the identification of suspected key species, i.e. animals that by their size, shape, abundance and/or activity might be important in regulating benthic community structure.

By definition, the baseline study ended when the pharmaceutical plant came into operation (March 1982). Thereafter, the work qualifies as a monitoring study. Within this ongoing phase, natural and introduced hard substrates (in the intertidal and sublittoral) have been monitored at biologically seasonal intervals for pattern in species settlement/community development, and the soft-bottom macrobenthic fauna and associated sediments have been recharacterised on an annual basis. Since 1987, the annual recharacterisations have included sediment profile imagery (SPI) as giving additional dimension to the returns from the more traditional grab-and-core sampling. The littoral study has been correspondingly increased; this now comprises: (a) detailed descriptions of accessible shores and the distribution of their dominant macroorganisms, along a wave exposure gradient, seawards from the harbour area, and (b) the population dynamics and reproductive biology of dominant faunal elements (i.e. representative barnacles, limpets, littorinids, the mussel *Mytilus edulis* and the dog whelk *Nucella lapillus*).

STUDY AREA

The study area covers approximately 6km^2 and encompasses the Kinsale Harbour–Bandon Estuary confluency. Depths range from 5m at the inner section of the harbour to 20m in its outer reaches. There is a relatively deep 'basin' in the harbour proper, with another depression adjacent to Charles Fort, at the head of the north–south running channel that links the inner harbour with the approaches. The eastern flank of this channel is deeper than that off the western shoreline, and depth increases gradually seawards. The littoral and sublittoral study sites/sampling locations, which have been variously occupied since the inception of the investigation, are shown in Fig. 1.

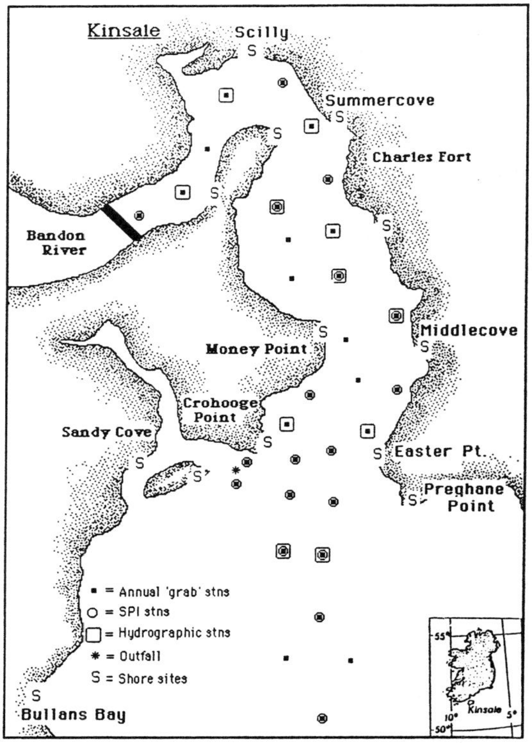

Fig. 1—Kinsale Harbour and approaches: representative sampling sites and stations, 1982–92.

Hydrography

The drainage from the River Bandon, which approaches Kinsale Harbour from the north-west, is a major determinant of hydrographic conditions in the area. Long term mean river flow is 13.53m^3 sec^{-1}, with an actual flow rate that rarely exceeds 50m^3 sec^{-1}. Predicted spring and neap mean tidal ranges at Kinsale are 3.5m and 1.8m respectively. Monthly hydrographic data from twelve stations across the Harbour show the water column to be quite variable; this variation results from seasonal effects of high or low runoff, the tidal state, and topographical effects of the coastline and the harbour bottom. The study area throughout clearly lies within the euhaline zone of the Bandon estuary; even lacking the salinity data, this could readily be inferred from the penetration of long-lived stenohaline species, such as the brittle-star *Amphiura filiformis*, up to the innermost sampling location. With respect to water temperatures, the classic estuarine situation obtains. Bottom readings are typically higher than those at the surface during the winter months, when there is also a gradient of increasing temperature from the inner to the outer harbour. During summer and autumn, this situation is reversed, with surface temperatures greater than off-bottom, and a decreasing gradient seawards. Dissolved oxygen levels range from 80–120% saturation—acceptable for estuarine waters.

Bedrock and sediments

Van Gelder and Clayton (1978) showed the study area to be characterised by a succession of predominantly muddy, non-calcareous strata known as the 'Kinsale Formation'. Where exposed, both in the littoral and sublittoral, the bedrock presents an aspect of slaty laminae of varying thickness. In the sublittoral, the cracks and crevices separating the laminae variously contain rock/shell debris, small stones and silt. Although the overall study area is characterised, for the most part, by finer deposits, 'pockets' of gravel do occur, with varying admixtures of biogenic and terrigenic elements. The inner harbour sediments are highly heterogeneous and, for the study area at large, exhibit the most variation. They tend to have high percentages of both silt-clay and gravel. Riverine discharge is assumed to be responsible for the silt-clay loading; the plant detritus present on and in these sediments also reflects the influence of the Bandon, particularly in times of heavy discharge. Variable freshwater input effects can cause bottom perturbation in the inner harbour, although its topography protects the area from the full force of storms.

The mid-western flank of the channel presents a 'belt' of clean, medium to fine sand, with very little silt or coarse material. This area is much shallower than the rest of the harbour and, hence, is more exposed to physical disturbance. The shallow depths and topography of the western shoreline suggest impedance of the outward flowing freshwater, which, interacting with the incursive marine flow, may cause small-scale gyres. These, allied with the probability that water flow is more rapid along this flank, would prevent the definitive deposition of silt and mud.

Sediment organic carbon levels have remained relatively constant across the study area over time, being typically <3% at most sampling stations, except for a few anomalously high readings. There tends to be a positive relationship between organic carbon and silt-clay levels, and the occasional high organic carbon readings are thought to be more attributable to the occasionally large amounts of autochthonous weed and terrestrially derived plant detritus that can accumulate in the inner harbour. Organic carbon tends to be generally lower in the outer harbour than in the inner harbour, suggesting a possible net dissipation of organics seawards. Sediment heavy metal levels have been assayed at various times during the course of the study and are typical for unpolluted estuaries/coastal waters.

Estuarial influences on sediment input, transportation and deposition are viewed as exerting a major control on animal distribution patterns. While never actually quantified, suspended detritus can be highly significant during and after periods of prolonged rainfall. One consequence is the attenuation of light at relatively shallow depths; this is thought to underlie the sparse macroalgal cover, which is a notable feature of the hard ground in the approaches to the harbour and, in turn, renders the area less than suited to herbivorous animals such as the sea-urchin *Echinus esculentus*. The *E. esculentus* population is composed of markedly smaller individuals than those of other Irish locations and White *et al.* (1985) were strongly of the opinion that this reflected an overall lack of quality food.

Exposure

Prevailing winds are from a southerly to a north-westerly direction. Direct observations have established that gales with a southerly aspect (more specifically from the 130–200°arc) can cause physical perturbation of the bottom sediments in all but the innermost parts of the harbour. A modelling exercise on the exposure of the study area confirmed the immediate outfall point as an area of high kinetic energy, i.e. a 'dispersive' location. This is also evidenced in the compact, clean, sandy ground of the area and is particularly highlighted in the findings of the 1987–91 SPI deployments.

Waste discharges

The position of the Eli Lilly outfall (by way of sea-bed diffusers) is shown in Fig. 1. Treated effluent from the Lilly plant is discharged from a sludge treatment facility. It is relatively high in salt ($NaCl^{-}$ <1%) and low in organic matter (COD≈ 200mgL). Other known waste inputs to the area are public body effluent and industrial effluent (mainly dairying) discharged into the Bandon from its catchment into the estuary at Kinsale.

FAUNA

Littoral

Sandycove Island lies close to the Eli Lilly outfall and was visited on a regular basis throughout the period of the investigation (1978–91). Transects were laid down on the north-east corner of the island and the findings for July 1991 (Fig. 2) serve not only to characterise the particular location but also to set the rocky intertidal of the area in broad perspective.

In July 1991, brown algae were represented by an abundance of *Lami-*

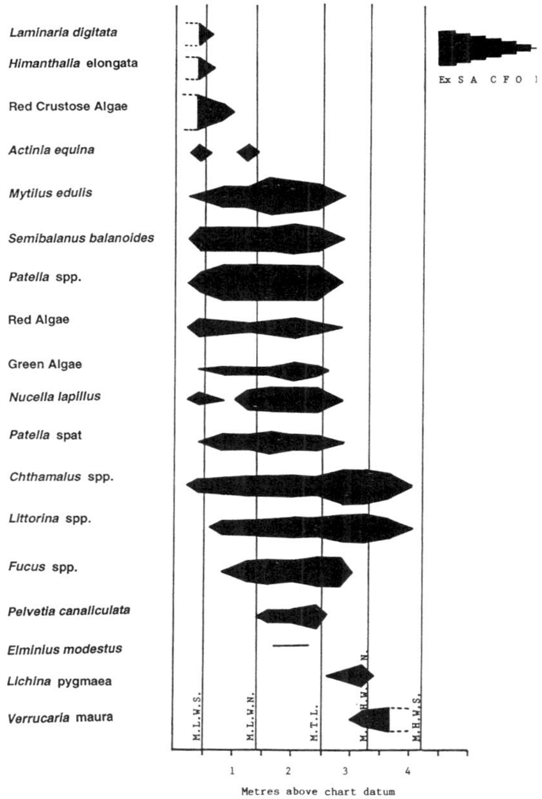

Fig. 2—Prominent elements in the littoral community at Sandycove Island, Kinsale Harbour, July 1991.

naria digitata, Himanthalia elongata and *Alaria esculenta* at mean low water of springs, and by fucoids (*Fucus serratus* and *Fucus vesiculosis*) and *Pelvetia canaliculata* in the mid-littoral region. Lichens (*Lichina pygmaea* and *Verrucaria maura*) were common on the upper shore. Green algae (*Ulva lactuca* mainly, with some *Enteromorpha* spp) were present only on patellid shells on the middle and lower shore. Red algae, located mostly in the *Mytilus edulis* beds, included *Porphyra umbilicalis, Ceramium* spp, *Lomentaria* spp, *Laurentia pinnatifida* and *Corallina officinalis.* Red crustose algae (*Lithothamnion* spp) were abundant in pools and on open bedrock at low tide. The bedrock, which consists of vertical ridges and hollows running up the shore, extends the vertical distribution of animal species such as anemones (*Actinia equina* and *Sagartia elegans*) and the dogwhelk *Nucella lapillus*, which occurred in the sheltered hollows but not on the exposed ridges. Barnacles (*Chthamalus* spp and *Semibalanus balanoides*), mussels (*Mytilus edulis*) and limpets (*Patella vulgata* and *Patella aspera*) were the dominant faunal elements, extending over most of the intertidal region. There was evidence of *Balanus balanoides* recruitment, while patellid spat and juveniles were abundant in crevices in the mid-shore region. Littorinids (mainly *Littorina neritoides* and *L. neglecta*) were densely aggregated in cracks and crevices on the upper shore. *Eulalia viridis* was recorded within the *Mytilus* beds on the middle shore, while sponges (*Halichondria panicea* and *Hymeniacidon perleve)*, bryozoans and the blue rayed limpet *Helcion pellucidum* (on *Laminaria digitata* fronds only) were taken on the lower shore. The occurrence of a *Pelvetia* zone, and other characteristics typically associated with more sheltered shores, such as the upshore extension in the distribution of dogwhelks (*Nucella lapillus*) and anemones (*Actinia equina* and *Sagartia elegans*), are attributed to the shelter proffered by the micro-topography of the bedrock. However, in the main, the shore is considered to be very exposed and is allocated an exposure rating of 2 on the Ballantine scale of 1–8 (Ballantine 1961).

Species composition along the transect was remarkably constant over the period 1978–91. While there were some minor differences, most of which were taken to reflect the slight difference in the times of sampling, there was no diminution in the prominence of any of the previously recorded species. *Elminius modestus,* an Australian barnacle species, first recorded in Britain in 1945 and not recorded in transects at Sandycove Island from 1978 to 1986, was observed in sparsely distributed clusters in 1989 and 1991.

Comparing all shores where multiple transects have been carried out, long-term studies suggest that the intertidal plant and animal communities of Kinsale Harbour are as can be expected under the types of 'natural' co-ditions obtaining in the area. Overall, there were no apparent signs of physical disturbance, e.g. the major cover species were well represented and perennial species exhibited broad size ranges from juveniles to adults. The shore at Easter Point, however, showed dramatic changes after the severe winter storms in 1989/90, when most of the mussel beds along the area of the transect were washed away. By June 1990 the area had been colonised by short lived, opportunistic species of green algae (*Enteromorpha* spp) and some *Fucus* spp. By the following summer (1991), very little green alga was observed along the line of transect, while fucoids were abundant. In August 1992, fucoids were still the dominant flora, but mussels (*Mytilus edulis*) were common, while limpets, both adults and spat, were extremely abundant on the middle and lower shore. The barnacle *Balanus balanoides* was common among mussel patches.

Imposex in Nucella lapillus

A study of nine sites across the area revealed that all populations of *Nucella lapillus* contain a percentage of animals showing the phenomenon of 'imposex', i.e. the induction of male sexual characteristics in the female. The phenomenon is at its most pronounced closest to the Kinsale marina and falls away towards the mouth of the harbour. Imposex in *N. lapillus* has been demonstrated by many investigators to be related to the presence of tributyltin. Tributyltin (TBT) is present in antifouling paints, which up to recently had widespread use on all marine vessels and fish enclosures in Ireland. Under new legislation, it cannot be used in freshwater (under any circumstances) or in the sea except on vessels over 25m long. Apparently, it was first introduced into Ireland in the late 1960s, and was undoubtedly in use in Kinsale before the Eli Lilly-funded studies began. TBT is only active as organotin (TBTO). In addition to the imposex factor, it has been shown to retard growth in oysters and clams, to cause shell distortion in oysters, to give rise to high levels of mortality in mussel larvae, to lessen the survival potential of certain other molluscs and crustaceans, and to produce an algistatic effect on diatoms and flagellates (see Smith and McVeagh 1991). Sensitivity to organotin varies between species and can be significant even at extremely low concentrations of the chemical. The situation is compounded by the fact that organotin is claimed to have a long residency time in marine sediments. What the wider implications are for the Kinsale fauna remains to be established. *Nucella lapillus* is still relatively abundant in Kinsale Harbour, as are most other species reported to be sensitive to this pollutant at some stage in their life histories.

Sublittoral rock

As revealed by extensive and regular diving surveys, coupled with the dredged returns from the 1978–81 characterisation, the 1979–82 baseline study, and occasional submersible-based video surveys, the epifauna of the study area can be broadly assigned to the lower *Laminaria hyperborea* association (i.e. a well-documented, hard-bottom community of plants and animals from European seas; see *inter alios* Könnecker and Keegan 1983). In the approaches to the harbour, the lower level of the *Laminaria hyperborea* forest lies at around 12m (i.e. a relatively shallow limit, which is presumed to reflect light attenuation by suspended seston in the water column), with some small and crustose algae reaching to 17m. The substrate over wide areas is covered by dense aggregations of the brittle-stars *Ophiothrix fragilis* and *Ophiocomina nigra,* with the crinoid *Antedon bifida*, and the sea-cucumber *Aslia lefevrei a*lso prominent. Owing to the frequently heavy layer of sediment, the epibenthic animals (with few exceptions) are confined to the vertical rock faces, crevices and the underside of stones. While the dense aggregations of brittle-stars inhibit the development of a prominent epifauna, the sponge *Cliona celata* (in its massive form), the hydroid *Sertularella polyzonias*, the octocoral *Alcyonium digitatum,* and large patches of the sea-mats *Parasmitina trispinosa* and *Schizomavella* spp are visually prominent on some vertical cliff faces. Within the depth range of the macroalgae, the hydroids *Alglaophenia pluma* and *Halecium halecinum* are also common. The large suspension-feeding polychaete worm *Bispira volutacornis* can be a dominating feature of the 'crevice' fauna. A particular feature of the epifauna is the low level of representation by the Tunicata. This has also been observed as a feature of the lower *Laminaria hyperborea* association in the outer Galway Bay area and is in marked contrast to the findings for inner Galway Bay (Könnecker 1981).

Settlement/colonisation patterns were followed on introduced surfaces over the period 1981–6. Some species settled out on the introduced substrates, which had not been found on 'natural' ground. This is cautiously interpreted as suggesting the absence of suitable terrain under natural conditions.

A study of the population dynamics of *Ophiothrix fragilis* has shown the aggregations to be composed mainly of adult and juvenile specimens with variable numbers of intermediate sized individuals (Ball 1991). Mean density was calculated at 1049.67 individuals/m^2, with a range of 363–3386 individuals/m^2. The predatory asteroid *Luidia ciliaris,* which is present in low numbers throughout the year, is found in large 'herds' during the summer months—98% of the *L. ciliaris* examined had their stomachs filled with *Ophiothrix fragilis* remains, with as many as 13 ophiuroids in a single specimen. There is no indication of optimal foraging, and no obvious selection of a particular prey size class, probably because of the large number of prey items available. Despite this predation, the population of *Ophiothrix fragilis* did not decline over the period 1981–6 and, if anything, actually increased in density.

Sublittoral sediments

The bottom sediments are colonised by a highly diverse, species-rich fauna, with a large number of individuals. The spatial diversity pattern over the area tends to reflect changes in dominance patterns rather than in the number of species. Against this, highest diversity values are found in the inner harbour reaches and are attributed to the heterogeneous or mixed nature of the associated sediments.

Four groups of stations (faunal assemblages) have been distinguished by classificatory analyses (Dinneen *et al.* 1986; Neiland 1992). As shown in Fig. 3, these are designated as Groups 1, 2a, 2b and 3, on the appreciation that 2a and 2b are more closely identified with each other than with any other group. These four assemblages do not form discrete units but intergrade and closely resemble each other. They differ quantitatively rather than qualitatively, sharing a strikingly large number of species that occur frequently, have a widespread distribution and are numerically important in the area as a whole. Changes in the abundances of the majority of these species, together with the presence of some less frequently occurring members, contribute significantly to the separation of the four assemblage groupings.

Comparing these faunal assemblages with near-equivalents (see Jones 1950; Thorson 1957) in temperate latitudes of the north-eastern Atlantic coastal zone, the Group 1 assemblage approximates an admixture of *Abra alba* and *Venus striatula* communities; Group 2a resembles the *Tellina fabula* faciation of the *Venus striatula* community; Group 2b approximates the classical *Venus striatula* community, and Group 3 is equivalent to the *Amphiura filiformis* community.

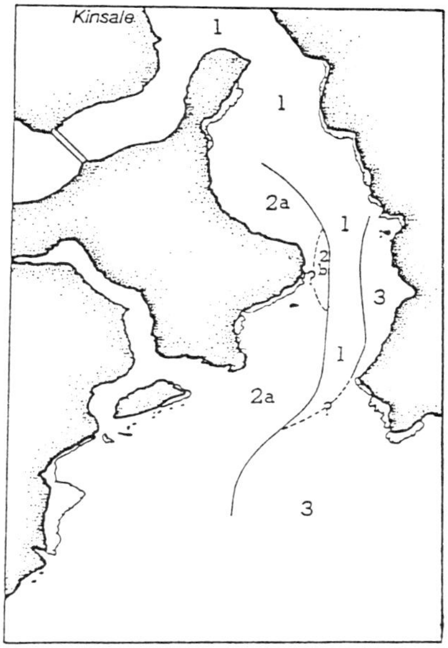

Fig. 3—The four infaunal assemblages delineated in the 1978–81 characterising survey and maintained through the ensuing annual recharacterisations (1982–92).

Because of their numerical prominence and presumed ecological importance, only the Polychaeta, Bivalvia, Gastropoda, Echinodermata and Crustacea have been taken into account in the annual recharacterisations. Analyses of the individual year and overall data sets, for both fauna and sediment, yield very similar results, revealing, on the one hand, the stable character of the macrobenthic infaunal communities and, on the other, the enduring correlation between the fauna and the nature of the deposit substrates. Three distinct sedimentary regimes can be identified: an admixture of coarse material with silt, in the basin constituting the inner harbour; a belt of clean, fine sand with very little coarse or silt material, extending down the mid-western flank of the 'channel'; and fine, silty deposits, extending along the channel's eastern flank and into the deeper harbour. The distribution of the faunal communities essentially reflects the same pattern: an assemblage characterised by the more detritivorous species, especially gastropods and crustaceans, and those favouring coarser, silty deposits in the inner harbour; suspension feeders dominating in the cleaner sand of the western flank of the channel; and the channel's eastern flank and the outer harbour being colonised by species typical of high successional stage, muddy-sand assemblages.

The data gleaned from the 1987–92 SPI images broadly support, and give additional dimension to, the more traditionally derived findings.

COMMENT

Littoral and benthic animals offer advantages in monitoring programmes; since many are sessile or relatively immobile, they must tolerate interference or die. They can thus integrate the effects of pollution over time and are considered as perhaps the best all round biological foci in marine monitoring protocols. Subtidal soft-bottom animals have further advantages in that their taxonomy is relatively easy, they can be sampled quantitatively with some degree of confidence, and there is an extensive literature on their community effects.

In terms of the littoral and benthic fauna of Kinsale Harbour, there is nothing to suggest that such minor changes as have been recorded since the inception of the programme in 1978 lie outside the system's inherent capacity for natural change. The assumption is made that the TBT effects were at their maxima, or had peaked, in the early stages of the study. The growing database continues to reinforce the early appreciation that dimensions of natural stress are imposed on the fauna by the often very high suspended load of the River Bandon, and by occasional turbation from river flow and gales with a southerly to easterly aspect. Severe climatic conditions can exaggerate these stresses and equally, and often very dramatically, bring about significant and visible change in the intertidal (e.g. the sloughing away of mussel patches to be replaced by incursive algae). Overall, the system clearly has a high measure of resilience. One positive side to the high kinetic energy effects is the basically 'dispersive' character of the area. The immediate outfall site can certainly be so designated, and bottom turbation has been registered at all but a small number of the sublittoral sampling locations.

ACKNOWLEDGEMENTS

The contribution of John Galvin and Albert Lawless (senior technicians at the Department of Zoology, MRI, UCG) warrants signal recognition; in their participation, and that of students and research scientists affiliated to the Zoology Department, resides the essential continuity of this study. The writers also applaud the enthusiasm and support of the management and technical staff of Eli Lilly (Irish Branch) S.A. for the 'Kinsale project'.

REFERENCES

BALL, B. 1991 Epibenthic studies in Kinsale Harbour, 1978–1987, with particular reference to the biology and ecology of a mixed aggregation of *Ophiothrix fragilis* and *Ophiocomina nigra* (Ophiuroidea: Echinodermata) in Kinsale Harbour (south coast of Ireland). Unpublished PhD thesis, University College, Galway.

BALLANTINE, W.J. 1961 A biologically-defined exposure scale for the comparative description of rocky shores. *Field Studies* 1(3), 1–19.

DINNEEN, P., KEEGAN, B. F., CROWE, W.A., O'CONNOR, B., McGRATH, D., COSTELLO, M. and KÖNNECKER, G. 1986 Littoral and benthic investigations on the south coast of Ireland—1. Macrofaunal distribution patterns and their relationship to the sedimentary environment in Kinsale Harbour. *Proceedings of the Royal Irish Academy* 86B, 81–106.

JONES, N. 1950 Marine bottom communities. *Biological Reviews* 25, 283–313.

KÖNNECKER, G. 1981 The epifauna of Galway Bay and Kilkieran Bay with special consideration of the Porifera, Hydroidea and Bryozoa. Unpublished PhD thesis, University College, Galway.

KÖNNECKER, G. and KEEGAN, B.F. 1983 Littoral and benthic investigations on the west coast of Ireland—XVIII. The epibenthic animal associations of Kilkieran Bay. *Proceedings of the Royal Irish Academy* 83B, 309–24.

NEILAND, S. 1992 Long-term studies on the benthic infauna of Kinsale Harbour with particular reference to the Polychaeta. Unpublished PhD thesis, University College, Galway.

SMITH, P.J. and McVEAGH, M. 1991 Widespread organotin pollution in New Zealand coastal waters as indicated by imposex in dogwhelks. *Marine Pollution Bulletin* 22, 409–13.

THORSON, G. 1957 Bottom communities (sublittoral or shallow shelf). *Memoirs of the Geological Society of America* 67, 46–534.

WHITE, M., KEEGAN, B.F. and LEAHY, Y. 1985 Growth retardation in the regular echinoid *Echinus esculentus* Linnaeus. In B.F. Keegan and B. O'Connor (eds), *Echinodermata 369–76.* Rotterdam. Balkema.

VAN GELDER, A. and CLAYTON, G. 1978 New data on early Dinantian (early Carboniferous) stratigraphy and sedimentation in south Cork (Ireland). *Geologie en Mignbrouw* 51, 25–32.

In: J.D. Reynolds (ed.) 1996 *The conservation of aquatic systems*, 80–83. Dublin. Royal Irish Academy.

OBLIGATIONS AND REQUIREMENTS FOR CONSERVATION OF THE IRISH AQUATIC BIOTA

Julian D. Reynolds

Department of Zoology, Trinity College, Dublin

ABSTRACT

Threatened Irish aquatic species fall into different categories: those that are widespread in Ireland but rarer elsewhere, such as the otter, those that are widespread elsewhere but rare in Ireland, such as the natterjack toad, and those that are uncommon everywhere, such as the freshwater pearl mussel. Our national and international obligations differ for each group, as will the best strategy for their protection. Conservation approaches must also recognise differing human values: there will be more support for protection of a higher vertebrate than for an alga or sponge. This can only be changed through education.

Some groups of Ireland's aquatic fauna and flora are depauperate compared with those of continental Europe, reflecting their dispersal mechanisms and island biogeographical processes and the relatively recent disruption of land and sea patterns resulting from glaciation. In other cases (e.g. sublittoral marine invertebrates) the biota is still insufficiently studied for it to be categorised comprehensively (but see Wilson and Lawler, this volume). Among aquatic or amphibious species Ireland has about a hundred vascular plant species (Curtis and McGough 1988; Scannell and Synnott 1987), perhaps 50 birds, not all breeding here, 30 freshwater fish, less than half of them native (Maitland, this volume), and 3000 invertebrate species (Illies, 1978). We might expect 5–20% of these to be threatened or endangered, using IUCN estimates.

While political boundaries may lead to the recognition of different threatened species and rarities in Northern Ireland and the Republic (see, for example, Curtis and McGough 1988; Stewart and Church 1992), we can categorise several classes of rare or threatened species, each with its particular conservation obligations and strategies (see Table 1).

1) Those that are widespread and possibly common in Ireland, but declining or rare elsewhere, e.g. otter (Lunnon, this volume; O'Sullivan, this volume), frog, freshwater crayfish (Matthews and Reynolds, this volume), salmon, the freshwater shrimp *Gammarus duebeni,* certain migratory waterfowl such as Greenland white-fronted geese and swans. Ireland is an important refugium for them in global terms.

(2) Those that are rare or declining in Ireland, but relatively secure elsewhere, e.g. the arctic char *Salvelinus alpinus*, the cave shrimp *Niphargus kochianus,* the natterjack toad *Bufo calamita,* certain aquatic insects, and other relict species from arctic, temperate continental or lusitanian biomes. In this category also belong certain introduced species, but these are not generally cherished until they have shown themselves to be a rare part of the biota for an extended time period, e.g. mute swans, or the sweet flag *Acorus calamus.*

(3) Those that are rare and/or declining throughout their range, e.g. the freshwater pearl mussel (Chesney, this volume); sea trout (Whelan and Poole, this volume); the sturgeon *Acipenser sturio.*

DISTRIBUTION AND THREATENED STATUS: CAUSATIVE FACTORS

Many marine species have pelagic larvae, ensuring a widespread distribution wherever habitats are suitable. Range curtailments or extensions may then follow slight changes in ocean currents or temperatures. To many freshwater species, however, the catchment or water body is sharply defined, and beyond which they may only rarely pass. It is vital to know not only the biogeography of the Irish aquatic species and the relative stability of their habitat, but something of their biology—longevity, fecundity, powers of dispersal, population structure and conditions for life.

Species that are widespread or common in Ireland but declining or rare elsewhere are so for a number of reasons, including the occurrence of particularly suitable environmental conditions here, such as temperature or dissolved oxygen; the scarcity or absence of certain epizootic diseases, such as the fungal plague *Aphanomyces astaci* afflicting freshwater crayfish, or levels of water pollution and restriction or alteration of habitat quality elsewhere.

Species rare or declining in Ireland but relatively healthy elsewhere may be relicts from arctic, temperate continental or lusitanian biomes, reflecting a once greater spread of climatic conditions, with past opportunities for colonisation. These may be at risk from large-scale changes such as global warming (see Grainger, this volume).

Species that are rare or declining everywhere are those whose niches have become altered or restricted throughout their ranges, often because of human impacts. For example, the freshwater pearl mussel *Margaritifera margaritifera* is a long-lived, large species that breeds late; sustained human fishing pressure or interference with the habitat may cause irreversible loss.

Table 1—Representative rare and threatened aquatic species in each status group (Group 1: widespread or common in Ireland, rarer elsewhere; Group 2: rare in Ireland, commoner elsewhere; Group 3: rare or declining throughout their range).

Taxon	*Group 1*	*Group 2*	*Group 3*
Tracheophyta	*Groenlandia densa*	*Hydrilla verticillata* *Ranunculus fluitans* *Potentilla fruticosa* *Hottonia palustris* *Viola persicifolia* *Puccinellia fasciculata*	*Pilularia globulifera* *Eriophorum gracile* *Hammarbya paludosa* *Zostera* spp
Porifera	*Homaxinella subdola*		
Coelenterata	*Pachycheiranthus multiplicatus* *Aptasia mutabilis* *Swiftia pallida* *Helcapoides elongatus*		
Crustacea	*Austropotamobius pallipes*	*Eurycercus glacialis* *Niphargus kochianus* *Mysis relicta* *Bathynectes longipes*	*Tanymastix stagnalis*
Mollusca			*Lymnaea glabra*
Teleostei	*Salmo salar*	*Salvelinus alpinus* *Coregonus autumnalis* *Gobius cruentatis*	*Acipenser sturio* *Alosa alosa*
Amphibia	*Rana temporaria*	*Bufo calamita* *Triturus vulgaris*	
Mammalia	*Lutra lutra*		
Aves	*Anser albifrons flavirostris*	*Cinclus cinclus*	

Exploitation may be a problem in each status group, depending on how the organism is collected (Tully, this volume). Trapping depends on an animal's behaviour, and economic forces dictate whether trapping continues once population is reduced; the presence of economically useful by-catch species is relevant here. Thus trapping freshwater crayfish will cease if numbers are uneconomic, while trapping for lobsters in the sea will continue because crabs are available as a by-catch. Fishing for pearl mussels may continue until stocks are nearly extinct, because of the value of the product.

CONSERVATION SOLUTIONS

There are varied approaches to successful conservation of the biota, depending in part on their rarity status in Ireland and elsewhere. Some conservation approaches are outlined in the keynote papers and posters following, e.g. Chesney; Marnell; Matthews and Reynolds; O'Sullivan; Whelan and Poole. In all cases, education and dissemination of information is important, but particularly for more obscure rare species.

Group 1 species that are widespread in Ireland but declining or rare elsewhere may have illogical conservation strategies imposed on them because of our international obligations. These may include bans on harvesting, e.g. frogs and crayfish. For such species careful monitoring and education may be a better policy than prohibition of capture.

For rare species or those with a restricted Irish distribution (group 2), our obligations are national rather than international. Conservation interest is often related to genetically isolated 'endemic' stocks, or to ecological/evolutionary conditions resulting in a perceived contrast between our tiny stocks and those elsewhere. Often the best approach is to conserve habitat, e.g. for the natterjack toad and char.

For species rare or in decline everywhere (group 3), we tend to follow international concerns in establishing conservation criteria unless a commercial or sporting interest is affected. Species of human economic interest may need protection by limiting exploitation, e.g. greenland white-fronted geese and lobster. Habitat conservation is usually recommended.

In summary, no single strategy is appropriate for the conservation of rare or declining aquatic species, because of their differing situations and our corresponding national and international obligations. Human value systems will tend to place higher priorities on certain flowering plants, birds, mammals and showy insects than on obscure invertebrates or lower plants. These attitudes can only be altered through education.

ACKNOWLEDGEMENTS

I wish to acknowledge the help given to me by my colleagues in providing information and in discussion of ideas.

REFERENCES

CURTIS, T.G.F. and McGOUGH, H.N. 1988 *The Irish red data book 1. Vascular plants.* Dublin. Stationery Office.

ILLIES, J. (ed.) 1978 *Limnofauna Europaea* (2nd edn). Stuttgart. Gustav Fischer.

SCANNELL, M.J.P. and SYNNOTT, D.M. 1987 *Census catalogue of the flora of Ireland.* (2nd edn). Dublin. Stationery Office.

STEWART, N.F. and CHURCH, J.M. 1992 *Red data books of Britain and Ireland: Stoneworts.* Peterborough. Joint Nature Conservation Committee.

In: J.D. Reynolds (ed.) 1996 *The conservation of aquatic systems*, 84–100. Dublin. Royal Irish Academy.

THREATENED FISHES OF THE BRITISH ISLES, WITH SPECIAL REFERENCE TO IRELAND

Peter S. Maitland

Fish Conservation Centre, Easter Cringate, Stirling, Scotland

ABSTRACT

Recent glaciation has resulted in temperate fish communities being much less diverse than those of the tropics and even in the British Isles there is a marked reduction in species from south to north. Recent checklists of the freshwater fish of Great Britain and Ireland have included 42 native and thirteen introduced species. Of the former, however, two are now extinct and eight are threatened in various ways, while several important stocks of common native species are under threat.

Human impact in Great Britain and Ireland has destroyed fish habitat on a wide scale and hundreds of populations have disappeared over the last two centuries. Protection for freshwater fish exists mainly through legislation and protected sites of various designations. The legislation is inadequate in some respects and very few of the protected sites were actually established for their fish interest.

The main conservation options for managing freshwater fish resources are habitat restoration, stock transfer to new sites, captive breeding and cryopreservation. The first two of these are regarded as the most useful for the long-term conservation of threatened species. Improved legislation would also help, especially in relation to preventing the import and transfer of potentially harmful species.

As well as the implementation of conservation management programmes for rare fish, research and monitoring studies are also needed in order to aid management and maintain a watch on populations of threatened species. In addition, national fish databases and registers of important stocks are an important background for overall fish conservation and management.

INTRODUCTION

Owing to the effect of periods of glaciation, temperate fish communities are much less diverse than those in the tropics, and even within the British Isles there is a marked reduction in species from south to north. For example, a river entering the north-west coast of Ireland is likely to have fewer than ten fish species, whereas an equivalent river entering the south-east coast of England may have over twenty species. Human impact has destroyed fish habitats in the British Isles on a wide scale and many hundreds of populations have disappeared over the last two centuries.

The recorded freshwater fish fauna of the British Isles consists of 42 native and thirteen introduced species, but the latter are not regarded as relevant in conservation terms (Table 1). Using objective criteria (Maitland and Lyle 1991) to assess the current status of native species, eight have been found to be under threat in various ways and two are now believed to be extinct here.

DEGREE OF THREAT

Various methods have been developed for assessing the conservation status of plant and animal species. The major classification system used internationally for assessing the status of threat is that adopted and developed by the International Union for the Conservation of Nature and Natural Resources (IUCN 1990). The *Red data book* categories are used by IUCN to indicate the degree of threat to individual species in their wild habitats. Below are given formal definitions of the categories together with additional information and examples to clarify and interpret them in relation to fish (Maitland 1986).

Extinct

This category is used only for species which are no longer known to exist in the wild after repeated searches of the type localities and other known or likely places. As interpreted by IUCN, this includes species which are extinct in the wild but surviving in captivity in aquaria and ponds.

Endangered

Defined as including taxa in danger of extinction and whose survival is unlikely if the causal factors continue operating, this group includes taxa whose numbers have been reduced to a critical level or whose habitats have been so drastically damaged that they are deemed to be in immediate danger of extinction. It may also cover species with populations so critically small that a breeding collapse through lack of genetic diversity becomes a possibility, whether or not they are threatened by human activities. An example would be a species reduced to a few specimens occurring in one inaccessible lake remote from human threats, but where one adverse event (e.g. a land-slide) could remove the whole population. In the case of fish, unlike most plants and other vertebrates, the actual numbers of individuals in the population are less important than the numbers of sites. This is because fish are often so confined within their habitat that one incident (e.g. a

Table 1—Freshwater fish recorded from England, Wales, Scotland and Ireland (N=Native, V=Vagrant, I=Introduced, E=Extinct).

	ENG	*WAL*	*SCO*	*IRL*
Sea lamprey *Petromyzon marinus* Linnaeus 1758	N	N	N	N
River lamprey *Lampetra fluviatilis* (Linnaeus 1758)	N	N	N	N
Brook lamprey *Lampetra planeri* (Bloch 1784)	N	N	N	N
Common sturgeon *Acipenser sturio* Linnaeus 1758	V	V	V	V
Allis shad *Alosa alosa* (Linnaeus 1758)	N	N	N	N
Twaite shad *Alosa fallax* (Lacepede 1803)	N	N	N	N
Atlantic salmon *Salmo salar* Linnaeus 1758	N	N	N	N
Brown trout *Salmo trutta* Linnaeus 1758	N	N	N	N
Pink salmon *Oncorhynchus gorbuscha* (Walbaum 1792)	V	–	V	–
Rainbow trout *Oncorhynchus mykiss* (Walbaum 1792)	I	I	I	I
Arctic charr *Salvelinus alpinus* (Linnaeus 1758)	N	N	N	N
Brook charr *Salvelinus fontinalis* (Mitchill 1815)	I	I	I	–
Houting *Coregonus oxyrinchus* (Linnaeus 1758)	E	–	–	–
Powan *Coregonus lavaretus* (Linnaeus 1758)	N	N	N	–
Pollan *Coregonus autumnalis* (Pallas 1776)	–	–	–	N
Vendace *Coregonus albula* (Linnaeus 1758)	N	–	E	–
Grayling *Thymallus thymallus* (Linnaeus 1758)	N	N	I	–
Smelt *Osmerus eperlanus* (Linnaeus 1758)	N	N	N	N
Pike *Esox lucius* Linnaeus 1758	N	N	N	N
Carp *Cyprinus carpio* Linnaeus 1758	I	I	I	I
Crucian carp *Carassius carassius* (Linnaeus 1758)	I	I	I	–
Goldfish *Carassius auratus* (Linnaeus 1758	I	I	I	–
Barbel *Barbus barbus* (Linnaeus 1758)	N	–	–	–
Gudgeon *Gobio gobio* (Linnaeus 1758)	N	N	I	N
Tench *Tinca tinca* (Linnaeus 1758)	N	N	I	I
Silver bream *Blicca bjoerkna* (Linnaeus 1758)	N	–	–	–
Common bream *Abramis brama* (Linnaeus 1758)	N	N	I	N
Bleak *Alburnus alburnus* (Linnaeus 1758)	N	–	–	–
Minnow *Phoxinus phoxinus* (Linnaeus 1758)	N	N	N	N
Bitterling *Rhodeus sericeus* (Bloch 1782)	I	–	–	–
Rudd *Scardinius erythrophthalmus* (Linnaeus 1758)	N	N	I	N
Roach *Rutilus rutilus* (Linnaeus 1758)	N	N	N	I
Chub *Leuciscus cephalus* (Linnaeus 1758)	N	N	N	–
Orfe *Leuciscus idus* (Linnaeus 1758)	I	I	I	–
Dace *Leuciscus leuciscus* (Linnaeus 1758)	N	N	I	I
Spined loach *Cobitis taenia* Linnaeus 1758	N	–	–	–
Stone loach *Noemacheilus barbatulus* (Linnaeus 1758)	N	N	N	N
European eel *Anguilla anguilla* (Linnaeus 1758)	N	N	N	N
Three-spined stickleback *Gasterosteus aculeatus* Linnaeus 1758	N	N	N	N
Ten-spined stickleback *Pungitius pungitius* (Linnaeus 1758)	N	N	N	N
Danube catfish *Silurus glanis* Linnaeus 1758	I	–	–	–
Burbot *Lota lota* (Linnaeus 1758)	E	–	–	–
Sea bass *Dicentrarchus labrax* (Linnaeus 1758)	N	N	N	N
Largemouth bass *Micropterus salmoides* (Lacepede 1802	I	–	–	–
Pumpkinseed *Lepomis gibbosus* (Linnaeus 1758)	I	–	–	–
Rock bass *Ambloplites rupestris* (Rafinesque-Schmaltz 1817)	I	–	–	–
Perch *Perca fluviatilis* Linnaeus 1758	N	N	N	N
Ruffe *Gymnocephalus cernua* (Linnaeus 1758)	N	N	I	–
Pikeperch *Stizostedion lucioperca* (Linnaeus 1758)	I	–	–	–
Common goby *Pomatoschistus microps* (Kroyer 1840)	N	N	N	N
Thick-lipped mullet *Chelon labrosus* (Risso 1826)	N	N	N	N
Thin-lipped mullet *Liza ramada* (Risso 1826)	N	N	N	N
Golden mullet *Liza aurata* (Risso 1810)	N	N	N	N
Bullhead *Cottus gobio* Linnaeus 1758	N	N	N	–
Flounder *Platichthys flesus* (Linnaeus 1758)	N	N	N	N
TOTAL NUMBER OF SPECIES	54	40	42	32

poisoning) could destroy every fish present. Thus in theory it would be much safer to have, say, 1000 fish in each of ten lakes than 100,000 or more in one lake.

Vulnerable

These are taxa believed likely to move into the Endangered category in the near future if the causal factors continue operating. Included here are taxa of which most or all of the populations are decreasing because of over-exploitation, extensive destruction of habitat or other environmental disturbance, as well as taxa with populations which have been seriously depleted and whose ultimate security is not yet assured, and those taxa with populations which are still abundant but are under threat from serious adverse factors throughout their range.

Rare

This category includes taxa with small world populations which are not at present Endangered or Vulnerable but are at risk. They are usually localised within restricted geographical areas or habitats or are thinly scattered over a more extensive range.

Indeterminate

This category includes taxa thought to be Extinct, Endangered, Vulnerable or Rare but where there is not enough information to say which of these four categories is appropriate.

THREATENED SPECIES

The biology and status of the ten most threatened species in Great Britain and Ireland have been reviewed by Maitland and Lyle (1991). Five are anadromous species which live mostly in the sea but spawn in fresh water, and five are purely freshwater species. Six of the species occur in Ireland, two only in Great Britain, and two are extinct. A brief note of the distribution and status of each threatened species follows.

Species occurring in Ireland

Sturgeon Acipenser sturio (*Linnaeus 1758*). This enormous fish is really only a vagrant to freshwaters in the British Isles, since it never breeds here and only very occasionally now does one ever venture into freshwater here. It regularly grows to lengths of 1–2m and has been known to reach 3m or more, with weights well over 200kg. The original distribution was much wider than at present. In the sea it occurred along the entire coastline of Europe from the North Cape to the Baltic, Mediterranean and Black Seas. It is now much scarcer, occurring only occasionally around the British Isles and breeding in a few European rivers, notably the Gironde, Guadalquivir and lower Danube. Usually a few are caught each year off British and Irish coasts by commercial netsmen.

The sturgeon is endangered throughout its range and is included in Appendix I of the Convention of International Trade in Endangered Species (CITES). The main threats to the species are overfishing in the lower reaches of its spawning rivers, obstruction of river migration by pollution, dams and weirs, and habitat destruction of spawning and nursery areas by pollution and engineering works. It is now protected in Great Britain by the Wildlife and Countryside Act (1981).

Allis shad Alosa alosa (*Linnaeus 1758*). This is the larger of the two British shads and commonly grows to about 30–50cm, exceptionally to 70cm. It is found along the coasts of western Europe from southern Norway to Spain and in the Mediterranean eastwards to northern Italy. It occurs mainly in shallow coastal waters and estuaries but during the spawning migration it penetrates the lower reaches of larger rivers.

This is a rare and vulnerable species which is much less common than formerly over its whole range, including Ireland (Went 1953; Gibson 1956). Although it was previously believed to breed in the Severn and other rivers, there is now no known spawning site in the British Isles. However, recent evidence suggests that there may be one in the Solway area. The principal threats to the Allis shad are obstruction of river migration by pollution, dams and weirs, habitat destruction of spawning areas by pollution and engineering works, and overfishing in estuaries. It is now protected in Great Britain by the Wildlife and Countryside Act (1981).

Twaite shad Alosa fallax (*Lacepede 1803*). Though both shads are large herring-like fishes, the Twaite shad is normally rather smaller than the Allis shad, adult fish averaging 25–40cm with a maximum of about 55cm. It occurs along most of the west coast of Europe, from southern Norway to the eastern Mediterranean Sea, and in the lower reaches of large accessible rivers along these coasts.

There are a number of extremely interesting non-migratory freshwater populations of this fish in a few of the larger European lakes such as Como, Garda, Iseo, Lugano and Maggiore. In Ireland, Lough Leane in Killarney has such a population, which appears to have been isolated here for thousands of years. It is known as the Goureen and was given subspecific recognition by Regan (1916) as *Alosa fallax killarnensis*. This fish is rather smaller than its marine based relatives and rarely grows longer than 25cm (Trewavas 1938; Twomey 1957).

The Twaite shad no longer breeds at most of its previous sites in the British Isles and is now much rarer than formerly throughout its range. Only a few spawning populations are known in the British Isles (e.g. the rivers Severn and Usk), but there are probably others (e.g. in the Solway and perhaps in Ireland (Bracken and Kennedy 1967). The main threats to the

Twaite shad are obstruction of river migration by pollution, weirs and dams, habitat destruction of spawning and nursery areas by pollution and engineering works, and overfishing.

Arctic charr Salvelinus alpinus (*Linnaeus 1758*). The adults are typically about 20–25cm long (85–170g), but in some lakes they may reach over 35cm and 600g. Some exceptionally large charr have been recorded recently from the vicinity of floating fish farm cages where their diet is boosted by waste food pellets. The Arctic charr is a holarctic species occurring all round the northern hemisphere and, in the northern parts of its range, forms mixed non-anadromous and anadromous populations (comparable to brown and sea trout) where the sea-going charr can reach a large size. In Great Britain and Ireland all populations are non-anadromous.

The number of populations in Europe has declined substantially. In the British Isles, although many populations in the north-west of Scotland (Maitland *et al.* 1984) and Ireland (Went 1971; Ferguson 1981) seem relatively secure, there has been a steady loss of stocks further south. The main threats to Arctic charr are from acidification, eutrophication and overfishing. An increasing interest in charr farming may give rise to problems for wild stocks.

Pollan Coregonus autumnalis (*Pallas 1776*). In Ireland, the pollan usually reaches an adult size of 30–35cm (350–450g), though occasionally larger specimens are recorded. In Canada, fish as long as 46cm (1.4kg) have been recorded, but weights of up to 2.5kg have been reported from Siberia. The pollan occurs in coastal areas and in the lower reaches of arctic rivers in north-eastern Europe, Asia and north-western North America—from the White Sea to Alaska and the North West Territories. Its occurrence in Ireland is surprising and well outside the rest of its world range. There are no logical explanations for its occurrence here. It is found in five large lakes: Lough Neagh, upper and lower Lough Erne, and the Shannon lakes Lough Ree and Lough Derg.

Until recently, the taxonomic status of this species was uncertain. Gasowska (1964), studying bones of pollan, suggested that it was *Coregonus autumnalis*—found nowhere else in western Europe (McPhail 1966). This suggestion was not fully accepted until electrophoretic techniques showed that the pollan does indeed appear to be the same fish as the Arctic cisco, *Coregonus autumnalis* (Ferguson 1974; Ferguson *et al.* 1978).

Although it is anadromous throughout most of its northern range, the Irish populations are all non-migratory and purely freshwater. However, for a few years after the establishment of a power station creating a barrier on the short River Erne, between lower Lough Erne and the estuary, pollan were taken in salmon nets fished in the estuary and an examination of these fish showed that their rate of growth had increased in the estuary (Twomey 1956). This incident would suggest that pollan still retain the potential to become, at least locally, an anadromous species. All the lakes in which it occurs are large and relatively clean and deep, although Lough Neagh and

Lough Derg have undergone considerable eutrophication in recent years (Wilson 1984; Bowman, this volume).

There is no evidence of any decline worldwide. All known populations in Ireland are still extant but there are only five of these. Moreover, there have been no records from Upper Lough Erne this century and even its continued existence in Lower Lough Erne is in some doubt; there are no recent records from Lough Ree (A. Ferguson, pers. comm.). The main threats to the pollan are from eutrophication, overfishing, and the introduction of alien fish species or diseases.

Smelt Osmerus eperlanus (*Linnaeus 1758*). The smelt, or sparling as it is called in Scotland, is a small- to medium-sized fish whose adult size varies greatly according to habitat. The normal range in length is 10–20cm, but the fish can sometimes reach 30cm. In general, fish from non-migratory freshwater populations are much smaller than those which have lived in the sea. The smelt occurs from southern Norway around the western coast of Europe (including the Baltic Sea) to north-west Spain. It occurs in coastal waters and estuaries and migrates into large clean rivers at spawning time. The species is tolerant of wide salinity changes and there are several non-migratory purely freshwater populations in large freshwater lakes in Finland, Sweden and Norway. The sole freshwater population in the British Isles, in Rostherne Mere in Cheshire, became extinct in the 1920s, probably as a result of eutrophication.

The smelt was once a common species in the British Isles and occurred in most larger British rivers from the Clyde and Tay southwards. In Ireland it occurs in the Shannon and Fergus.Over the last century, the species has gone into decline and has disappeared from many rivers. The only notable exception to this is the River Thames where, although it did disappear for many decades, fish have returned naturally as pollution decreased. More recently, smelt have started returning to the Forth Estuary, where they have not been seen for over twenty years. The main threats to the smelt are pollution and overfishing, obstruction of migration by pollution, dams and weirs, and habitat destruction of spawning and nursery areas by pollution and engineering works.

Species not occurring in Ireland

Houting Coregonus oxyrinchus (*Linnaeus 1758*). The normal adult length is 25–35cm. This is the only truly anadromous whitefish in European waters and its numbers have undergone a severe decline during this century. It is primarily a fish of the eastern North Sea; in the British Isles it has only been known as a vagrant, but at one time it was taken quite regularly along the south-east coast of England. No specimens have been seen in Great Britain for many decades.

Thus the houting is now extremely rare and endangered throughout its range and is extinct in many places, including the British Isles. The main threats to this species are obstruction of river migration by pollution, dams and weirs, overfishing in estuaries and lower reaches of rivers, and habitat destruction of spawning and nursery areas by pollution and engineering works.

Powan Coregonus lavaretus (*Linnaeus 1758*). The adult size of powan is normally 30–35cm (300–400g), but in some European lakes fish up to 70cm (8kg) have been recorded. This species is widespread across much of north-west Russia, Finland and Sweden; it occurs also in Norway, Switzerland, Germany, Poland and France. In the British Isles it is a relatively rare local fish occurring in Loch Lomond and Loch Eck (Scotland), in Ullswater, Brotherswater, Haweswater and Red Tarn (England) and in Llyn Tegid (Wales).

The species has declined in several parts of Europe. In Great Britain the seven known populations are still extant, though very vulnerable. The main concerns arise from eutrophication, pollution and the introduction of new species of fish. It is now protected in Great Britain by the Wildlife and Countryside Act (1981).

Vendace Coregonus albula (*Linnaeus 1758*). The vendace is usually regarded as a small-to medium-sized fish with a normal adult size of 20–25cm (100–150g). It occurs in north-west Europe, from northern Scandinavia and north-west Russia in the north to Bavaria further south, and from the English Lake District in the west to western Russia in the east. In the British Isles it has been known from only four lakes, two in Scotland and two in England (Maitland 1970).

The species has declined across its range. The populations in Scotland are now extinct as a result of eutrophication. In England, the species occurs in larger waters—Bassenthwaite Lake and Derwentwater—and, though it may be secure there in the short term, this is at present Britain's rarest fresh-water fish and its status is of urgent concern. The main threats to the vendace are from eutrophication and the introduction of alien fish species and diseases. It is now protected in Great Britain by the Wildlife and Countryside Act (1981).

Burbot Lota lota (*Linnaeus 1758*). Adult burbot are usually 30–60cm long. The species has a northern circumpolar distribution and is found in clean lakes and rivers throughout much of northern, west and central Europe and in northern Asia and North America. In the British Isles, its original distribution was restricted to the larger rivers of eastern England but has not been seen since 1970 and is believed to be extinct. The main threats have been from pollution, river engineering and habitat destruction. It is protected in Great Britain by the Wildlife and Countryside Act (1981).

OTHER FISH STOCKS OF IMPORTANCE

Recently, ecologists and conservationists have paid more and more attention to the importance of the stock concept. Because of the isolation of many freshwater systems and the strong homing instinct at spawning time in migratory fish species, many populations in the British Isles appear to have evolved a stock individuality over the last few thousand years (Cross and Ward 1980; Ferguson 1981; Taggart *et al.* 1981). In some species (e.g. Atlantic salmon *Salmo salar* Linnaeus 1758 and brown trout *Salmo trutta* Linnaeus 1758) it is likely that this is being substantially interfered with through indiscriminate stocking or other careless management practices. However, there are still important stocks of these and several other species which are of particular note and worthy of special conservation measures.

Thus, as well as rare species, rare or unusual stocks of fish also deserve conservation attention. Several of these are considered below, but it is likely that others will be found worthy of attention once further research has been carried out on the stock individuality of other freshwater fish species in the British Isles.

The river lamprey *Lampetra fluviatilis* (Linnaeus 1758) is quite widespread and common in many running waters in Ireland and in Great Britain south of the Great Glen. However, in the Loch Lomond catchment there is a unique population (Maitland 1980), which differs in many respects from all others. Instead of migrating to the sea to feed on fish after metamorphosis, the stock of river lampreys in the River Endrick migrates downstream in the spring to Loch Lomond, where it feeds intensively on powan. In the autumn, it stops feeding and migrates back into the River Endrick, where it spawns in the spring. As well as having these different ecological characteristics, this stock is quite different morphologically from other river lampreys (Morris 1989): the Lomond lampreys are smaller, have larger oral discs, snouts and eyes and very much darker pigmentation than lampreys from other waters. There are also significant differences in body proportions and in the mean numbers of myomeres and teeth. These differences may result from both genetic isolation and the short freshwater feeding stage in Loch Lomond.

The brook lamprey *Lampetra planeri* (Bloch 1784) is the most widespread of the three lamprey species in Great Britain and Ireland, and because it does not migrate to the sea there are many more individual isolated stocks than is the case with the sea lamprey *Petromyzon marinus* Linnaeus 1758 or the river lamprey. Though the brook lamprey is clearly the smallest species at full maturity, it is the largest at metamorphosis (after which it does not feed), and most adults are in the range 110–170mm. However, there is one isolated population on the Isle of Skye which matures at a very much smaller size (80–90mm) than anywhere else in the British Isles.

The brown trout was formerly regarded as just one of several species of trout in the British Isles and about 50 species have been described throughout the brown trout's range. However, recent studies have indicated that all these fish belong to just one variable species—*Salmo trutta*. Nevertheless there is no doubt that substantial stock individuality exists within this species, though undoubtedly some stocks have been lost owing to pollution and other factors, including mismanagement by anglers. Some clearly defined stocks, such as ferox (*Salmo trutta ferox*), gillaroo (*Salmo trutta stomachius*) and sonaghan (*Salmo trutta nigripinnis*), live together in Lough Melvin in Ireland. Ferguson and Mason (1981) have shown that these stocks are genetically distinct and, though living together in the same water, spawn separately in different parts of the system, thus continuing and reinforcing reproductive isolation. Other important stocks occur in some of the remoter parts of Scotland (Stephen 1984). It is essential that all such stocks are clearly identified and protected from the careless stocking procedures practised by fishery managers in the past.

Unusual stocks of the widespread and common three-spined stickleback *Gasterosteus aculeatus* Linnaeus 1758 have been found in the Outer Hebrides by Campbell (1985). These fish, referred to as *anomalus* morphs, do not possess pelvic spines or pelvic girdles and have a variable number of dorsal spines. Four populations of these spine-deficient sticklebacks are now known from peatland lochs in North Uist and at least one of these stocks is made up wholly of the *anomalus* morph. In one population, some individuals have no spines at all—dorsal, pelvic or ventral. This morph occurs only rarely throughout the world range of the three-spined stickleback but other populations have been recorded along the Pacific coast of North America. Breeding experiments carried out by Campbell (1985) suggest that spine deficiency is not a recessive, easily masked, characteristic and that these stocks of *anomalus* are probably quite stable genetically.

THREATENED FISH COMMUNITIES

With a wide variety of freshwater habitats and 42 indigenous fish species of varied distribution there is potential for a considerable range of fish communities in Great Britain and Ireland which are worthy of conservation measures. Many of the original communities have been eliminated by pollution or habitat destruction and others have been altered by the introduction of foreign species. However, there are still valuable communities left, reflecting pristineness, simplicity, diversity, uniqueness and classical community structures. In addition, it should be emphasised that there are many freshwater habitats which have no fish and are of conservation importance (for invertebrate communities or other factors) because of this.

Naturally, the communities in all the waters with populations of threatened species are of special interest and importance. The value of several of these is further enhanced by the combination of various other native species, as discussed below. The question of how to identify such sites has been considered by Maitland (1985).

Original pristine fish communities are under substantial threat—indeed there are probably very few left. However, there are a fair number of rivers and lakes with classical oligotrophic native fish communities, especially in northern and western Ireland and Scotland.

There are many waters in the British Isles which naturally have very simple communities—sometimes with only one species present. Small streams or lakes in which only brown trout or three-spined sticklebacks occur are common in some coastal or highland areas and thus are not rare or unusual in themselves. However, some simple communities are of special interest and one such is Loch Meallt in Skye, where the only fish present are Arctic charr and three-spined sticklebacks (Campbell 1984). These species interact with each other in various ways and this site is of substantial conservation interest.

Some of the larger waters in the British Isles have a very diverse array of fish species, which make up extremely interesting and complex communities. Southern lowland rivers are prominent here, notably the Rivers Avon, Great Ouse, Severn and Thames in England. The larger lake communities are less complex, but notable diversity occurs in Llyn Tegid, Loch Lomond, Loughs Corrib and Erne. All these waters contain almost full representations of the original stenohaline fishes although in some waters these are now well mixed with later introductions. Such communities should be kept under observation and if possible safeguarded.

Other waters can be of particular interest because of unusual, sometimes unique, combinations of species. This is especially true where threatened species or unusual races are involved. Good examples of this are Loch Eck and Haweswater, the only two sites in the British Isles where Arctic charr and powan occur together, Loch Lomond where powan are found with river lamprey, and Llyn Tegid where powan (Gwyniad) occur with grayling *Thymallus thymallus* (Linnaeus 1758). Lough Leane's uniqueness in having a freshwater population of Twaite shad (goureen) is further enhanced by the presence there of Arctic charr.

In giving full consideration to the conservation value of any site, intactness is of importance. As just noted, Loch Eck and Haweswater are the only sites in the British Isles with both Arctic charr and powan. However, whilst the integrity of Haweswater has been substantially compromised by the fact that it has been dammed and is now a fluctuating water supply reservoir, Loch Eck is relatively intact. In addition, Loch Eck has the full range of British salmonid species, including the migratory Atlantic salmon and sea trout, neither of which occur in Haweswater. Loch Eck is clearly the more important site in conservation terms.

Provisional lists of some important fish communities in the British Isles have been tabulated by Maitland and Lyle (1991). It should be remembered that while none of the native fish species is endemic—they all occur elsewhere in north-western Europe—they have mostly been isolated from their common stock since the end of the last ice age and some have developed into distinct local races. This differs, say, from the great majority of our native birds, whose populations are, in the main, continuous with those of mainland Europe.

As discussed above, the main human pressures on native fish communities are those resulting from exploitation, pollution, loss of habitat and the irresponsible re-distribution of disruptive fish species from other parts of the country. In recent years there has been a marked trend in the British Isles away from natural mixed freshwater fish populations towards artificially maintained unstable stocks of a few species for sporting and commercial interests. However, it should be noted that all ten of the threatened fish species considered above were once of commercial importance in the British Isles and, in a few places, some still are.

PROTECTING FISH

The identification of these threatened fishes and important sites for them has provided a foundation for a review of existing protection in Great Britain and Ireland and for a major fish conservation management programme (Maitland and Lyle 1992).

Current protection for freshwater fish in Great Britain and Ireland exists mainly through legislation, but this is inadequate in some respects and tends to focus mainly on sport and commercial species (Maitland and Campbell 1992; Maitland and Lyle 1992). There is also a large number of protected areas, of various designations, for native flora and fauna, but very few have been established because of their fish. A recent survey of freshwaters and fish in 235 National Nature Reserves (NNRs) throughout Great Britain (Lyle and Maitland 1992) revealed that no NNRs have been declared for fish interests; 75% of NNRs included freshwaters, but of these 20% had no fish. Of the 51 fish species currently believed to occur in Great Britain—two of the 55 species listed in Table 1 are extinct, two are vagrants and one occurs only in Ireland—37 (73%) were recorded in NNRs during the survey. As might be expected, the majority of these fish are common species and there is little immediate concern for their conservation.

Of the ten species under threat only Arctic charr, powan and smelt are present in British NNRs. Although a number of reserves have diverse fish communities, only six are regarded as having particular conservation value. Outstanding among these is Loch Lomond NNR because of its high fish species diversity (fifteen native and four introduced), rare species (powan), a unique race (river lamprey) and unusual species combinations. There appears to be no comparable information from Ireland on the status of fish in waters protected for conservation.

The main future options for fish conservation in Great Britain and Ireland are habitat management and restoration, stock transfer to new sites, captive breeding and cryopreservation (Maitland and Lyle 1990; 1992). The first two of these are the most useful for the long-term conservation of threatened species. Improved legislation would also help, especially in relation to preventing the import and transfer of potentially harmful species (ICES 1973). The recent production of a Red Data Book for Irish vertebrates (Whilde1993) is an important step forward for the conservation of Irish fish.

Habitat restoration and management are major goals in the conservation of most species and communities. Obviously enormous damage has been done to many fish habitats and the situation is often not easy to reverse, especially in the short term, where fish species or communities are severely threatened. In many cases, potentially unique stocks have completely disappeared. Even where habitat restoration is contemplated, stock transfer (discussed below) could be an important interim measure. However, there are a number of important examples of habitat restoration in temperate areas and it should be emphasised that habitat restoration, protection and management are the principal long-term means through which successful fish conservation will be achieved.

There have been enormous advances in pollution control in the British Isles over the last few decades and a number of the worst rivers are now much cleaner. Thus, the rivers Clyde and Thames are now far less polluted than 50 years ago and fish have been returning to them in increasing numbers. At their worst, both rivers were virtually fishless in their lower reaches. Over the last two decades, many freshwater and estuarine species have returned to the lower Thames, which now supports a diverse community not unlike its original one. Rehabilitation of the River Clyde has been rather slower but salmonid fish are conclusive evidence of high water quality, and the return of the Atlantic salmon to this river after an absence of more than 100 years is a tribute to decades of work by the local river purification board.

There are many other examples of habitat restoration to both running and standing waters. These include the control of eutrophication in some lakes, the neutralising of some important acidified waters, the reinstatement of spawning gravels, the restoration of meanders where rivers have been canalised, the provision of more compensation water where over-abstraction has taken place, etc. However, very few of these examples relate directly to the restoration of threatened fish populations *per se*. The types of habitat 'improvement' carried out by many anglers (e.g. deepening of pools, clearance of riparian trees, etc.) are really to help angling rather than the aquatic habitat.

As well as promoting the importance of habitat management and conservation at specific sites, Maitland and Lyle (1990) have implemented trial programmes of translocation for Arctic charr, powan and vendace to create and safeguard new populations. Specific criteria have been developed in planning such translocations. For example, the translocation activities must pose no threat to the parent stock, nor to the ecology or scientific interest of the introduction site, which clearly must be ecologically suitable. In general, sites from which the species concerned has disappeared should be considered unsuitable unless the causal factors have been identified and removed. Ideally, the introduction site should be in the same catchment or the same geographic region as the parent stock; or in the same geographic region as a former stock, now extinct. Permission must be obtained from riparian owners, relevant statutory authorities, or conservation agencies in the case of nature reserves. Stock may be transferred as eggs, fry, juveniles or adults, but the removal of adults could pose a threat to the parent stock. Special consideration should be given to the genetic integrity of the stock to be translocated. Once the stock has been defined, maximum genetic diversity should be sought by selecting material widely in space and time. The transfer of undesirable diseases or parasites can be avoided by taking eggs only from the parent stock and checking for disease. Notes of each translocation experiment should be kept and details published where relevant. Finally, it is important to monitor the fate of the translocated stock.

As well as the implementation of conservation management programmes for rare fish, research and monitoring studies are also needed to aid such management and to maintain an awareness of the status of important species.

CONCLUSIONS

Internationally, relatively little attention has been paid to the conservation of fish but, fortunately, this attitude has changed in recent years and there is now an increasing interest in this topic in most parts of the world (e.g. McDowall 1983; Pollard *et al.* 1990). In North America, in particular, there have been several important initiatives (Johnson and Rinne 1982; Williams and Miller 1990), including the creation of a facility devoted entirely to the captive breeding and rearing of rare species. It is hoped that the recent review of fish conservation in Great Britain and Ireland (Maitland and Lyle 1992) has provided a firm foundation for the conservation of fish in the British Isles and it is hoped that the proposals emanating from it will be implemented in future years. The main conclusions from the review are as follows.

1. Further legislation to control the import of potentially disruptive fish is required.
2. Further controls are needed on the movement and release of fish within Great Britain and Ireland.
3. Fishery management practices must be improved and there should be a complete ban on live baiting.
4. Active conservation management is needed for seven species (Allis shad, twaite shad, Arctic charr, powan, vendace, pollan and smelt).
5. Further international support should be given to the conservation of three other species in Europe (sturgeon, houting and burbot). Two of these (sturgeon and houting) are helped by their listing in Annex IV of the Habitats and Species Directive of the European Community, but they should also be protected in other parts of Europe where they occur.
6. Management plans should be established for all major sites of threatened species.
7. Habitat restoration and management should be the major long-term objective, but it is often slow and expensive in the short term.
8. Translocation of threatened stocks to create new safeguard populations is an important measure, perhaps vital in the short term (e.g. for vendace) and valuable in the long term.
9. Regular monitoring of existing sites and of new sites is essential (Kallio-Nyberg and Koljonen 1991).
10. Research is still required on several species and is especially urgent for some (e.g. Allis shad).

11. Substantial resources are required to implement conservation proposals, which must involve several different organisations in Great Britain and Ireland.
12. Once safeguarded, there is no reason to discourage wise sustainable use of any stocks (e.g. for angling or commercial exploitation).
13. As well as legislation and active conservation management there should be a programme to educate the public (especially anglers) about the value of native fishes.
14. There should be regular reviews of the distribution and status of freshwater fish, and development and maintenance of registers of important stocks.

The general conclusion is that, although there has previously been some legislation and management in relation both to fish and various general aspects of conservation (such as the establishment of nature reserves), little of this has been aimed directly at the protection of threatened fish species. This situation must be improved if further valuable stocks of native species are not to be lost. However, there is every reason to hope that, given adequate support, threatened fish species, unusual races and communities and, indeed, fish stocks in general can be managed more positively, thereby saving for posterity this valuable and renewable resource, so important for scientific, recreational, commercial and aesthetic purposes.

ACKNOWLEDGEMENTS

Much of the work described was funded by the Nature Conservancy Council. I am grateful to Dr J.J. Bracken, Dr A. Ferguson and Dr A. Whilde for information on Irish fish and to Dr J.D. Reynolds for helpful comments on a draft of this paper.

REFERENCES

BRACKEN, J. and KENNEDY, M. 1967 Notes on some Irish estuarine and inshore fishes. *Irish Fisheries Investigations* 3B, 4–8.

CAMPBELL, R.N. 1985 Morphological variation in the three-spined stickleback (*Gasterosteus aculeatus*) in Scotland. *Behaviour* 93, 161–8.

CAMPBELL, R.N. 1984 Predation by the Arctic charr on the three-spined stickleback and its nest in Loch Meallt, Skye. *Glasgow Naturalist* 20, 409–13.

CROSS, T.F. and WARD, R.D. 1980 Protein variation and duplicate loci in the Atlantic salmon, *Salmo salar* L. *Genetics Research, Cambridge* 36, 147–65.

FERGUSON, A. 1974 The genetic relationships of the coregonid fishes of Great Britain and Ireland indicated by electrophoretic analysis of tissue proteins. *Journal of Fish Biology* 6, 311–15.

FERGUSON, A. 1981 Systematics of Irish charr as indicated by electrophoretic analysis of tissue proteins. *Biochemical Systematic Ecology* 9, 225–32.

FERGUSON, A., HIMBERG, K.J.M. and SVARDSON, G. 1978 The systematics of the Irish pollan (*Coregonus pollan* Thompson): an electrophoretic comparison with other holarctic Coregoninae. *Journal of Fish Biology* 12, 221–33.

FERGUSON, A. and MASON, F.M. 1981 Allozyme evidence for reproductively isolated sympatric populations of brown trout *Salmo trutta* L. in Lough Melvin, Ireland. *Journal of Fish Biology* 18, 628–42.

GASOWSKA, M. 1964 Coregonids classification discussed on the basis of *Coregonus pollan* Thompson from Lough Neagh (Northern Ireland). *Annales Zoologici, Warszawa* 22, 413–19.

GIBSON, F.A. 1956 Further specimens of shads from Irish waters. *Irish Naturalists' Journal* 12, 109.

ICES 1973 Code of practice to reduce the risks of adverse effects arising from the introduction of non-indigenous marine species. *Conseil International pour l'Exploration de la Mer*, 50–1.

IUCN 1990 *The red data book of endangered animals*. Gland. International Union for the Conservation of Nature and Natural Resources.

JOHNSON, J.E. and RINNE, J.N. 1982 The Endangered Species Act and southwestern fishes. *Fisheries* 7, 2–7.

KALLIO-NYBERG, I. and KOLJONEN, M. 1991 The Finnish char (*Salvelinus alpinus*) stock register. *Finnish Fisheries Research* 12, 77–82.

LYLE, A.A. and MAITLAND, P.S. 1992 Conservation of freshwater fish in the British Isles: the status of fish in National Nature Reserves. *Aquatic Conservation: Marine and Freshwater Ecosystems* 2, 19–34.

McDOWALL, R.M. 1983 New Zealand's freshwater fish: their conservation status and conservation need. *Pacific Science Congress Proceedings* 15, 153–60.

McPHAIL, J.D. 1966 The *Coregonus autumnalis* complex in Alaska and northwestern Canada. *Journal of the Fisheries Research Board of Canada* 23, 141–8.

MAITLAND, P.S. 1970 The origin and present distribution of *Coregonus* in the British Isles. In C.C. Lindsey and C.S. Woods (eds), *Biology of coregonid fishes,* 99–114. Winnipeg. University of Manitoba Press.

MAITLAND, P.S. 1980 Scarring of whitefish (*Coregonus lavaretus*) by European river lamprey (*Lampetra fluviatilis*) in Loch Lomond, Scotland. *Canadian Journal of Fisheries and Aquatic Sciences* 37, 1981–8.

MAITLAND, P.S. 1985 Criteria for the selection of important sites for freshwater fish in the British Isles. *Biological Conservation* 31, 335–53.

MAITLAND, P.S. 1986 *Conservation of threatened freshwater fish in Europe*. Strasbourg. Council of Europe.

MAITLAND, P.S. and CAMPBELL, R.N. 1992 *Freshwater fishes of the British Isles*. London. Harper Collins.

MAITLAND, P.S., GREER, R.B., CAMPBELL, R.N. and FRIEND, G.F.

1984 The status and biology of Arctic charr, *Salvelinus alpinus* L., in Scotland. In L. Johnson and B. Burns (eds), *Biology of the Arctic charr: Proceedings of the International Symposium on Arctic charr*, 193–215. Winnipeg. University of Manitoba Press.

MAITLAND, P.S. and LYLE, A.A. 1990 Practical conservation of British fishes: current action on six declining species. *Journal of Fish Biology* 37A, 255–6.

MAITLAND, P.S. and LYLE, A.A. 1991 Conservation of freshwater fish in the British Isles: the status and biology of threatened species. *Aquatic Conservation, Marine and Freshwater Ecosystems* 1, 25–54.

MAITLAND, P.S. and LYLE, A.A. 1992 Conservation of freshwater fish in the British Isles: proposals for management. *Aquatic Conservation, Marine and Freshwater Ecosystems* 2, 165–83.

MORRIS, K.H. 1989 A multivariate morphometric and meristic description of a population of freshwater-feeding river lampreys, *Lampetra fluviatilis* (L.), from Loch Lomond. *Zoological Journal of the Linnaean Society* 89, 357–71.

POLLARD, D.A., INGRAM, B.A., HARRIS, J.H. and REYNOLDS, L.F. 1990 Threatened fishes in Australia—an overview. *Journal of Fish Biology* 37A, 67–78.

REGAN, C.T. 1916 The British fishes of the subfamily Clupeinae and related species in other seas. *Annals and Magazine of Natural History* 18, 1–18.

STEPHEN, A.B. 1984 Electrophoretic evidence for population variation in brown trout (*Salmo trutta*) and the implications for management. In A. Holden (ed.), 119–27. *Proceedings of the 15th Annual Study Course*. Stirling. Institute of Fisheries Management.

TAGGART, J., FERGUSON, A. and MASON, F.M. 1981 Genetic variation in Irish populations of brown trout (*Salmo trutta* L.): electrophoretic analysis of allozymes. *Comparative Biochemistry and Physiology* 69B, 393–412.

TREWAVAS, E. 1938 The Killarney shad or goureen (*Alosa fallax killarnensis*). *Proceedings of the Linnaean Society* 150, 110–12.

TWOMEY, E. 1956 Pollan of Lough Erne. *Irish Naturalists' Journal* 12, 14–17.

TWOMEY, E. 1957 A specimen of shad *Alosa fallax killarnensis* from Lough Leane, Killarney. *Irish Naturalists' Journal* 12, 270.

WENT, A.E.J. 1953 The status of the shads, *Alosa finta* and *Alosa alosa* (Cuvier), in Irish waters. *Irish Naturalists' Journal* 11, 8–11.

WENT, A.E.J. 1971 The distribution of Irish char, *Salvelinus alpinus*. *Irish Fisheries Investigations* 6, 5–11.

WHILDE, A. 1993 *Irish red data book 2: Vertebrates*. Belfast. HMSO.

WILLIAMS, J.E. and MILLER, R.R. 1990 Conservation status of the North American fish fauna in fresh water. *Journal of Fish Biology* 37A, 79–85.

WILSON, J.P.F. 1984 The food of the pollan *Coregonus autumnalis pollan* Thompson in Lough Neagh, Northern Ireland. *Journal of Fish Biology* 23, 253–62.

In: J.D. Reynolds (ed.) 1996 *The conservation of aquatic systems*, 101–110. Dublin. Royal Irish Academy.

THE SEA TROUT STOCK COLLAPSE, 1989–92

K.F. Whelan and W.R. Poole

The Salmon Research Agency of Ireland, Farran Laboratory, Furnace, Newport, Co. Mayo

ABSTRACT

The sea trout (*Salmo trutta* L.) collapse, which affected many fisheries in the mid-west during the period 1989–92, is described. Particular emphasis is placed on sea trout data from the Burrishoole catchment, where quantitative estimates of survival are available. A general review of all research data relating to the collapse is discussed as a case history of a species requiring conservation in a limited portion of its range. The overall response to the collapse is assessed in terms of: response time, the effectiveness of the actions taken, financing the response and future actions required to protect and enhance the remaining stocks.

INTRODUCTION

The sea trout (*Salmo trutta* L.) has been described as displaying an archetypal life history. It is considered the true original of the trout species, its ancestors having colonised freshwater some 10,000 years ago.

Adult sea trout return from the sea during the late spring and summer to spawn in the late autumn in tiny streams and rivulets located high in the mountains. Irish west coast sea trout are relatively slow growing and long lived, in comparison with salmon, and their life cycle is far more complex (Fahy 1985; Whelan 1989). Sea trout remain in freshwater for two to four years before migrating to sea as smolts and they will often return three or more times to spawn in freshwater.

Sea trout are essentially fish of acid waters (pH 5.5–6.5). In general, sea trout populations flourish where growth rates are poor, where survival in freshwater is difficult, and where there is easy access to the sea. Survival fluctuates widely, both in freshwater and at sea, and alternating periods of dearth and abundance may appear in consecutive years.

DECLINING STOCKS

Because of the marginal existence of the species, there has been increasing evidence during the past two decades of a slow decline in stocks from some catchments. This was largely attributed to poaching and a range of environmental problems such as field drainage, stream drainage and maintenance, fertilisation of the hillsides, afforestation and, more recently, hillside erosion due to overgrazing by sheep (Whelan 1992).

Whelan (1991; 1992; 1993) has described the appearance of a more serious decline, which appeared in many fisheries along the western seaboard in 1986 and had, by 1989, resulted in a population collapse in many mid-western sea trout fisheries (Table 1). The major collapse occurred in 1989 when, tragically, there was little sea trout research taking place. Only anecdotal information is available on the sequence of events that took place during the May/June period of 1989. However, the following details are known.

- In May large numbers of post-smolts and some mending kelts appeared in the estuaries of the Delphi and Erriff systems.
- Many of these fish were heavily infested with lice and photographic evidence is available showing the level of lice damage to the skin and fins of the more seriously affected fish.

Table 1—Annual sea trout catches from mid-western fisheries, 1985–92.

Fishery	*1985*	*1986*	*1987*	*1988*	*1989*	*1990**	*1991**	*1992**
Kylemore	2411	1099	543	1116	198	10	450	200
Culfin	298	173	222	235	36	8	175	60
Erriff	770	433	450	308	120	60	235	293
Delphi	2150	1281	832	675	309	112	437	494
Carrowniskey	135	97	90	160	45	N/R	65	10
Bunowen	475	110	146	340	88	10	120	25
Clifden R.	95	70	98	20	4	N/R	6	N/R
Belclare	95	70	98	400	0	6	60	20
Ardbear	86	75	36	27	5	N/R	15	N/R
Newport	1155	1485	783	1049	135	51	268	30
Burrishoole	497	614	237	245	41	39	106	25
Inagh	N/R	N/R	1369	824	29	10	7	45
Inver Beg	254	220	67	18	0	0	10	50
Ballynahinch Upr.	378	398	306	173	10	0	0	N/R
Ballynahinch Mid.	202	150	224	75	5	0	30	45
Ballynahinch Lwr.	2300	2000	1500	850	20	90	200	50
Gowla	1035	967	266	210	0	0	0	0
Doohulla	200	150	100	20	1	N/R	N/R	N/R
Crumlin	328	222	261	26	0	N/R	N/R	30
Invermore	1481	1345	325	199	48	0	0	0
Carna	60	180	100	60	3	N/R	N/R	N/R
Costello and Fermoyle	2745	2316	1698	1851	462	140	234	375
Letermuckoo	74	50	100	30	2	0	0	N/R

* 1990–2: All sea trout taken on a catch and release basis

* 1992: Provisional figures

- In some fisheries, such as Delphi and Kylemore, appreciable numbers of very thin adult fish appeared during the angling season and the majority of these displayed poor gonadal development. Not all of these fish showed evidence of lice damage. Thin fish that did not display lice damage also appeared in the Crumlin and Costello/Fermoyle fisheries.

In 1990 the Salmon Research Agency co-ordinated a broadly based research programme which, it was hoped, would identify the extent of the problem and define a possible cause or causes. The results of this programme are summarised in Whelan (1992).

- In the Burrishoole catchment there was evidence of a long-term decline in sea trout stocks, overlaid by a severe population collapse during the period 1987–1990, particularly during 1989–90.
- The most severe collapse in sea trout stock levels occurred in the area from Galway Bay in the north to Clew Bay in the south.
- In the so-called 'affected area' there was evidence of a significant drop in growth at sea: finnock were particularly small and maiden sea trout were no larger than normal finnock. There were also changes in growth in freshwater, with larger than normal smolts migrating to sea.
- Many smolts returned to freshwater after spending only a matter of weeks at sea. These were heavily infested with juvenile sea lice (*Lepeophtheirus salmonis).* Levels as high as 200 were recorded on an individual 20cm fish.
- Lice had caused severe skin and flesh damage to the trout. Observational evidence indicated that some of these fish later contracted secondary infections and died.
- No evidence was found of a disease consistent with the stock collapse.
- Many of the more fanciful theories that were rampant during the initial stages of the research were discounted. There was no evidence of sea trout feeding in the vicinity of cages, where it was claimed pollack and coalfish were eating the fish.
- From the stomachs of marine fish caught while sampling for sea trout, there was no evidence of reduced food availability at sea or a lack of specific food items (e.g. sand-eels).
- A detailed study of historical records of water temperature and water height showed that subtle interactions between both factors resulted in the successful exodus of smolts. Two critical temperatures were suggested: a temperature of 7–8°C must be reached before exodus commences, while recruitment was adversely affected by the occurrence of temperatures of 13°C or higher, before migration had taken place.

The obvious manifestation of the problem affecting the sea trout was severe sea lice infestation, but the preliminary research did not answer the vital questions as to the source of the lice and whether or not there was some other physiological or disease factor predisposing the fish to the unprecedented sea lice infestation. It was, however, hypothesised by the

Sea Trout Action Group (STAG) that a major factor in the collapse of sea trout stocks during 1989 and 1990 was a sea lice population explosion deriving from coastal salmon farms. The group also accepted that physiological stress, disease and environmental factors may have played a role in both the long-term decline and the collapse of 1989 and 1990 (Anon. 1991a).

FURTHER RESEARCH

During the following two years the research programme concentrated on five principal areas:

1. Monitoring of sea trout survival in a range of fisheries,
2. Further investigations into the role of sea lice,
3. A physiological profile of sea trout smolts to study their state of fitness when entering saltwater,
4. Disease monitoring,
5. The development of practical sea trout enhancement strategies.

In 1991 the Department of the Marine established a Sea Trout Working Group and the results of the sea trout research programme were examined by this group in December 1991 and 1992 (Anon. 1991b; 1993).

Whelan (1993) discusses the possible role played by such factors as increased acidity, physiological stress, disease, and increased parasite loadings (both external and internal) on sea trout survival. In addition the Sea Trout Working Group also considered these factors in the context of the sea trout stock collapse (Anon. 1993) and reached the following conclusions.

- No evidence of disease was observed in sea trout examined during 1992. A variety of internal parasites was observed and identified but these did not provide an explanation for the collapse of sea trout stocks.
- While acid flushes or chronic acidification may be contributory factors in the sea trout problem in some catchments, there is no direct evidence to identify a causal link.
- A number of man-made environmental changes have taken place and these undoubtedly affect stocks in specific catchments. The sea trout stock collapse, however, cannot be attributed either individually or collectively to these changes.
- No global failure of osmoregulatory ability of sea trout, which could explain the premature return of sea trout smolts to rivers, was identified.
- Sea lice infestations on sea trout continued to be a major problem in 1992. Data examined by the Working Group in 1992 again failed to show a causal link between sea lice on farmed salmon and infestations on sea trout. A relationship was postulated, however, between production of juvenile sea lice larvae from fish farms and the mean intensity of sea lice found on prematurely returning sea trout during May. Failure to find the free living larval stages severely hampered attempts to investigate the presence or absence of a causal link between the production of sea lice larvae and infestations of sea lice on sea trout.

The only consistent factors, therefore, to emerge from the research carried out to date are the premature return of both smolts and kelts to the estuaries in early to late May, and the presence of abnormal numbers of juvenile lice. A number of additional key points have also emerged from the sea lice research (Anon. 1991b; 1993; Tully and Whelan 1993a; Tully *et al.* 1993a; Tully *et al.* 1993b).

- In the western region during 1991, 95% of the total nauplius larval production of *Lepeophtheirus salmonis* was from farmed salmon.
- The smolts and kelts were infested with juvenile lice within two to three weeks of migrating to sea.
- The morphological and physiological impact of the lice on sea trout was significant and sufficient to cause mortality.
- Fish caught in estuaries sharing the same embayment generally had similar infestation levels.
- There was a suggested correlation in both April 1991 and 1992 between larval production from the farms and subsequent infestation levels on the sea trout three weeks later.
- The production of larvae in Killary Harbour in spring 1992 fell to zero owing to fallowing of the salmon farm, and the infestation levels on sea trout was subsequently reduced by 75%.
- Daily larval production in Clew Bay increased from zero in 1991 (farms fallow) to significant levels in May 1992 with the reintroduction of the salmon farms, and the subsequent levels of lice on sea trout increased from a median of 5.5 per fish in 1991 to over 55 in 1992.
- There was evidence in 1991, and particularly 1992, of sea trout problems in the south-west. Significant numbers of prematurely returning post-smolts and adult sea trout, heavily infested with sea lice, were recorded from the estuaries of the Inny R. and L. Currane.

SEA TROUT SURVIVAL IN THE BURRISHOOLE SYSTEM

Sea trout survival, both before and during the period of the stock collapse, has been quantitatively assessed at the Agency's traps (Fig. 1). The survival of smolt to first return as finnock in the same year has historically ranged from 11.4% to 32.4% (Fig. 2). In 1988 the survival fell below the previous recorded minimum to 8.5% and in 1989 to a minimum of 1.5%. There were increases in 1990 and 1991 to 5.1% and 10.0% respectively (Whelan 1993). However, in 1992, the survival fell to 3.7%, the second lowest marine survival recorded to date (Anon. 1993).

These events closely follow the presence or absence of salmon farms in inner Clew Bay. Salmon farming commenced in Clew Bay in the mid-1980s, reaching a peak in 1989 when three farms were in operation. Because of disease problems the bay was fallowed from April 1990 until January 1992 when 50,000 salmon, 1kg+, were towed into inner Clew Bay.

Data from trapping sites in the Gowla and Invermore systems over the

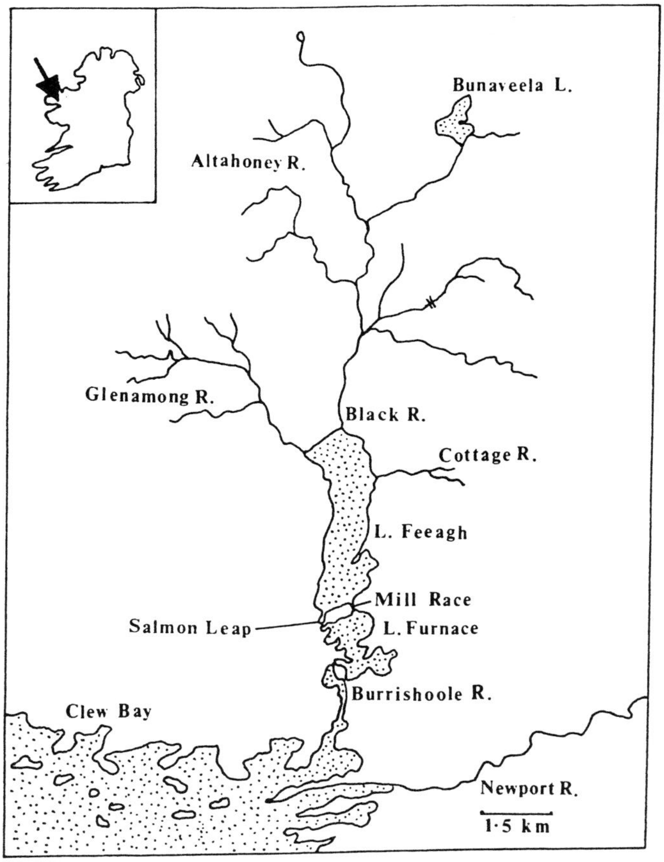

Fig. 1—The Burrishoole Fishery, Co. Mayo.

past two seasons have also indicated an unprecedented sea trout stock collapse in these systems (Anon. 1993).

DISCUSSION

Although many of the research results relating to the sea trout stock collapse have been available in report form over the past three years, to date no critical analysis of the problem and the overall response to the problem has been attempted.

It is our belief that the scale of the sea trout problem and the long-term implications for the stocks and their attendant fisheries are only now generally being realised. Perhaps it is timely to review the course of the problem and to draw from it the lessons to be learned in order to deal more effectively with similar ecological disasters in the future.

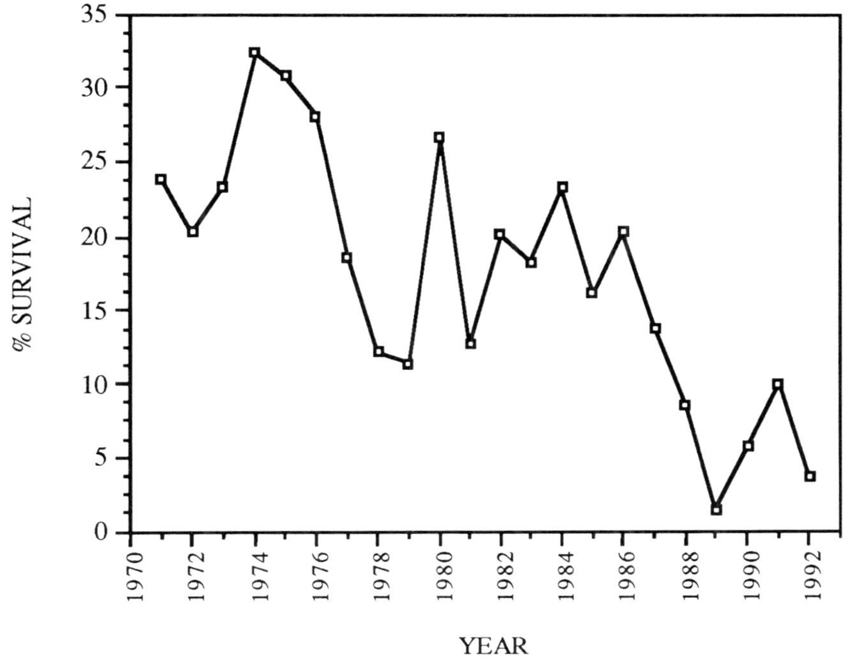

Fig. 2—Annual marine survival of smolts to finnock in the Burrishoole system.

Basic to the initial ineffective response to the problem in 1989 and 1990 was ignorance. Little was known of the complex biology of the sea trout in either the freshwater or marine environments; even less was known about the biology of the parasite that was manifestly a major component of the problem.

Sea trout research has been on-going for many years but, in the main, it has been poorly funded and was often carried out as a by-product of salmon research. In Ireland the species is not generally considered of true economic importance since it does not support a full commercial fishery. However, in terms of angling tourism it plays a pivotal role and its presence, even in a mixed salmon and sea trout fishery, is an integral part of the 'west of Ireland angling experience'.

While the extent of the research effort may have been confined over the years, we were fortunate in that the quality of the work carried out was excellent. This provided a sound basis on which to assess the relative age structure of the various stock components and, in the case of the Burrishoole fishery, provided census data on the survival of both smolts and the older age classes of sea trout. However, we know little of the population dynamics of the stocks and are still unable to answer the vital question: 'What makes a sea trout?' The trout's life at sea was a mystery, and we could only speculate on the factors affecting marine survival.

Our ignorance of the biology of the sea-louse was even more profound and to date it has proved impossible to locate in any quantity, in the wild, the first free-living larval stage (*Nauplius* I). Without a method to locate and quantify this life stage of the sea-louse it would seem difficult if not impossible to demonstrate conclusively a causal link between the production of juvenile lice from sea-farms and subsequent infestation levels on sea trout.

As outlined previously, the major sea trout stock collapse occurred in 1989, at a time when little sea trout research was taking place. When STAG was formalised in the autumn of that year one of the first tasks undertaken was to compile a list of possible causative factors or events leading to the sea trout collapse. Some eighteen separate factors were listed and chief amongst these was a sharp decline in elver numbers (known to be a food of sea trout) and an increase in seal numbers. STAG also decided that a broadly based research programme was required to address the problem and to define more clearly the factors involved in the stock collapse.

It is important to stress that STAG was a voluntary group encompassing all of the groups, agencies and individuals with an interest in the welfare of the west coast sea trout. It is also worth noting that the original group had no official representative from the fish farming industry since, at that time, no evidence existed to link fish-farming activities with the sea trout stock collapse. As a voluntary group STAG had no research funds available to it, nor were emergency funds available from central government. STAG therefore launched a major and very successful fund-raising campaign to which the Department of the Marine and many hundreds of individuals and groups subscribed.

The publication in report form of the results from the first year's research and the subsequent publication of the STAG Reports (Anon. 1991a; 1993), in which it was hypothesised for the first time that sea lice emanating from salmon farms were a major factor in the sea trout collapse, were pivotal to subsequent events. STAG was hoping for action based on the precautionary principle, to deal in a co-ordinated fashion with the lice problem on salmon farms. The group was very conscious of the critical stock situation in many mid-western sea trout fisheries—in addition to the loss of the adult stocks in 1989, the majority of fisheries had lost two year classes of post-smolts in 1989 and 1990.

It was argued that by dealing with the lice problem the salmon-farming industry would benefit and that the sea trout would also show improved survival. If STAG were proved wrong, then the salmon-farming industry would have gained the benefits of reduced lice infestations. Unfortunately, the preliminary research data available were judged insufficient to warrant a fully co-ordinated lice control campaign and, as outlined previously, it was recommended that an expanded research programme should be undertaken to consider the four areas of physiology, lice, disease, and environmental factors. The results from this programme were to be assessed by the Department of the Marine's Sea trout Working Group. It was also agreed that conservation measures were required to protect the remaining stocks of sea trout in the mid-west and bye-laws were introduced banning the killing of sea trout both in the commercial and sport fisheries.

The Salmon Research Agency has consistently argued since the publication of its first sea trout report in 1990 that enhancement of sea trout stocks should be considered a priority area (Salmon Research Agency, unpublished). Conscious of the long life-cycle of the sea trout and the lack of appropriate enhancement techniques, the Agency sought sponsorship for a novel kelt and finnock reconditioning programme. Funding was not available from either STAG or government sources but, thanks to the support of the Electricity Supply Board and BP Nutrition, this programme has been funded since the spring of 1991. It has proved very successful (Poole *et al.* 1994), and some 110,000 ova were stripped from 93 re-conditioned sea trout in December 1992 and January 1993.

Considerable efforts were made to co-ordinate the management of bays containing fish farms and to promote the concepts of single-bay management and fallowing. Such initiatives, if implemented speedily and effectively, should reduce organic loadings and help to control disease and lice problems on salmon farms.

It has also been agreed that a National Sea Trout Enhancement Programme will be initiated and that central to this programme will be the techniques of sea trout kelt re-conditioning and the on-rearing of sea trout juveniles in sea-cages. Initially, four separate stocks will be available to the programme. Their progeny will ensure the availability of a native 'living gene bank' with which to re-seed the affected fisheries in the mid-west.

The Report of the Sea Trout Working Group (Anon. 1993) has also recommended that the existing bye-laws, banning the killing of sea trout taken either on rod and line or in the draft-net fishery and restricting the areas and seasons for commercial drift-netting, should remain in place.

What then are the lessons to be learned from these four traumatic years? Firstly, we have a graphic demonstration of the need for basic research in advance of a serious ecological problem. There is also a clear need for an emergency fund to deal with such problems. We strongly urge the establishment of an ECO-FUND, which would assist with the mitigation of damage arising from ecological disasters such as the sea trout stock collapse. Management of the ECO-FUND should be assigned to a small expert group of biologists and administrators. The situation where donations from abroad are required to fund emergency research and protection programmes for native Irish species should be avoided.

Finally, we suggest that the difficulties and risks of applying the precautionary principle in all situations where a species is at risk must be accepted, without qualification, whether or not that species is of perceived economic value.

REFERENCES

ANON. 1991a The Sea Trout Action Group Report. *Sea Trout News* 2. Galway. Western Regional Fisheries Board.

ANON. 1991b *Report of the Sea Trout Working Group*. Dublin. Fisheries Research Centre, Department of the Marine.

ANON. 1993 *Report of the Sea Trout Working Group II*. Dublin. Department of the Marine.

FAHY, E. 1985 *Child of the tides*. Dublin. Glendale Press.

POOLE, W.R., DILLANE, M.G. and WHELAN, K.F. 1994 Artificial reconditioning of wild sea trout (*Salmo trutta* L.) as an enhancement option: initial results on growth and spawning success. *Fisheries Management and Ecology* 1 (3), 179–92.

TULLY, O. and WHELAN, K.F. 1993 Production of nauplii of *Lepeophtheirus salmonis* (Kroyer) (Copepoda: Caligidae) from farmed and wild Atlantic salmon and its relation to infestation of wild sea trout (*Salmo trutta* L.) off the west coast of Ireland in 1991. *Fisheries Research* 17, 187–200.

TULLY, O., POOLE, W.R. and WHELAN, K.F. 1993a Infestation parameters for *Lepeophtheirus salmonis* (Kroyer) (Copepoda: Caligidae) parasitic on sea trout (*Salmo trutta* L.) post smolts on the west coast of Ireland during 1990 and 1991. *Aquaculture and Fisheries Management* 24, 545–55.

TULLY O., POOLE, W.R., WHELAN, K.F. and MERIGOUX, S. 1993b Parameters and possible causes of epizootics of *Lepeophtheirus salmonis* (Kroyer) parasitic on sea trout (*Salmo trutta* L.) on the west coast of Ireland. In G.A. Boxhall and D. Defaye (eds), *Proceedings of the First European Crustacea Conference, Paris, August 31–September 5,* 199, 202–13. London. Ellis Horwood Series in Aquaculture and Fisheries Support.

WHELAN, K.F. 1989 *The angler in Ireland: game, coarse and sea*. Dublin. Country House.

WHELAN, K.F. 1991 Disappearing sea trout—decline or collapse? *The Salmon Net* 23, 24–31.

WHELAN, K.F. 1992 Management of salmon and sea trout stocks. In J. Feehan (ed.), *Environment and development in Ireland: Proceedings of a conference held at University College Dublin,* 457–66. Dublin. The Environmental Institute, University College Dublin.

WHELAN, K.F. 1993 Decline of sea trout in the west of Ireland: an indication of forthcoming marine problems for salmon? In D. Mills (ed.), *Proceedings of the Fourth International Atlantic Salmon Symposium, St Andrews, N.B. Canada, June 1992*, chapter 9. Oxford. Fishing News Books, Blackwell Scientific Publications.

In: J.D. Reynolds (ed.) 1996 *The conservation of aquatic systems*, 111–116. Dublin. Royal Irish Academy.

OTTER (*LUTRA LUTRA* L.) DISTRIBUTION IN IRELAND

Ruth Lunnon

Department of Zoology, Trinity College, Dublin

ABSTRACT

Between January 1990 and November 1991 a survey of otter signs in a range of freshwater and marine habitats across Ireland was carried out. The sites investigated (268) were selected from the sites visited (2373) in the 1980/81 National Otter Survey of Ireland, carried out by the Vincent Wildlife Trust. Data are presented on the relationship between various habitat quality factors and otter distribution, and comparisons are made between the distribution of 1980/81 and 1990/91, including an examination of the contrast in status of the otter in the Greater Dublin area and in the country as a whole.

INTRODUCTION

The European otter *Lutra lutra* L. is an elusive mustelid whose numbers have been in sharp decline in many parts of Western Europe since the late 1950s (Chanin and Jefferies 1978; Mason and MacDonald 1986; Jefferies 1989; Foster-Turley *et al.* 1990). In 1976, concerned nature conservation bodies in Britain set up the Joint Otter Group (JOG), which devised a standard methodology for the surveying of otter signs (O'Connor *et al.* 1977).

National baseline surveys were carried out in the UK in 1978–9 (Crawford *et al.* 1979; Green and Green 1980; Lenton *et al.* 1980) and later in Ireland in 1980–1 (Chapman and Chapman 1982). Alternate 50km squares of the Irish National Grid were surveyed and otter signs were found in 91.7% of the 2373 sites visited (Chapman and Chapman 1982).

Follow up surveys were completed in Britain in 1984–5 (Andrews and Crawford 1986; Green and Green 1987; Strachan *et al.* 1990) and, as no full-scale resurvey was planned for Ireland, it was decided in 1989 to visit a subsample of the 1980–1 baseline sites in order to assess the current distribution status of otters in Ireland.

METHOD

The methods used in the 1980–1 survey were followed (Chapman and Chapman 1982). The standard methodology, mentioned above, entailed a search for otter signs of a maximum of 600m of habitat. The search was terminated as soon as otter signs were found. Sites were spaced at 5-8km intervals along aquatic habitats, with approximately five sites being examined per 10km square of the National Grid. At each site a record of habitat details was made on a field sheet. Signs of the American mink *Mustela vison* (Schreber) were also noted when they were encountered.

A subsample of three 10km squares per 50km square was selected for resurveying. Thus, some 268 (11.3%) of the original survey sites were visited. The main survey was carried out between January 1990 and September 1991 and was conducted by the author, Wildlife Rangers employed by the Office of Public Works, and volunteers from the Irish Wildlife Federation and the Ulster Wildlife Trust. All those involved either had previous experience of otter survey work or received some training in the field. In addition, all 71 sites in the Dublin 50km square were visited by the author between November 1989 and November 1991.

RESULTS

Of the 268 sites surveyed, 240 (89.6%) contained otter signs (Fig. 1). The otter distribution pattern of 1980–1 (Fig. 2) remained unchanged in 64.2% of the 10km squares visited. Table 1 gives the results of chi-square analysis of otter distribution against some habitat factors (as assessed by surveyors). The relationship between distribution and water pollution (freshwater sites) was significant (x^2=13.7, $p<0.001$; 1 d.f.). The relationship between mink signs and the presence or absence of otters was not significant (Table 1). Mink scats were recorded at 11.9% of sites surveyed. This is likely to be an under-recording of their presence as, in most cases, spraints were found within a few metres of the start of a survey and thus no further investigation of the site was made.

In the Dublin 50km square (O.SW), out of 71 sites visited the number of locations positive for otters increased from 44 in 1980–1 (Chapman and Chapman 1982) to 48 in the present study. The distribution pattern of 1980–1 remained unchanged in just 25% of the 10km squares visited. Pollution was evident at over half (52.9%) of the sites where otters were absent, but the relationship was not significant. However, chi-square was significant for otter distribution and bankside cover ($x^2 = 4.4$; $p < 0.05$, 1 d.f.). A new record of otter signs was made on the River Liffey in the centre of Dublin City in November 1991.

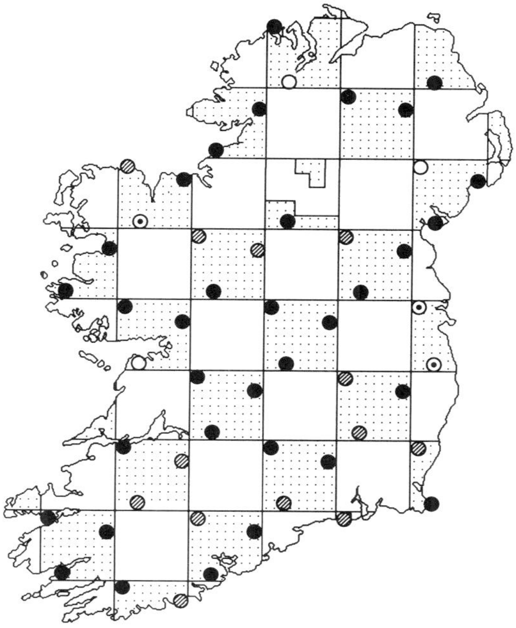

Fig. 1—Otter distribution records for 1990–1 showing the alternate 50km squares surveyed and details for the 1990–1 sites. Black circles: otters encountered at all sites in square; stippled: one site negative; circles with dot: two negative sites; white circles: three or more negative sites.

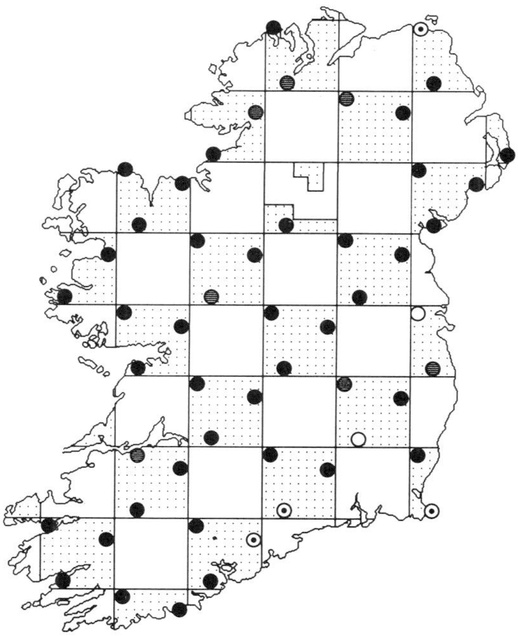

Fig. 2—Otter distribution records for 1980–1 (after Chapman and Chapman 1982).

Table 1—Numbers of sites that were positive or negative for otters in relation to four habitat-quality factors (as recorded on survey field sheets): chi-square test results.

A. Human disturbance

	High	*Low*
Positive	77	163
Negative	3	25

Chi-square = 0.87, n.s.; 1 d.f.

C. Mink

	Present	*Absent*
Positive	29	211
Negative	3	25

Chi-square = 0.01, n.s.; 1 d.f. (with Yates's Correction)

B. Bankside cover

	High	*Low*
Positive	123	117
Negative	13	15

Chi-square = 2.33, n.s.; 1 d.f.

D. Pollution—inland

	Polluted	*Unpolluted*
Positive	45	171
Negative	14	12

Chi-square = 13.7, $p < 0.0001$; 1 d.f.

DISCUSSION

The use of spraint surveys for monitoring otter distribution has been the subject of some debate (Jefferies 1986; Conroy and French 1987; Kruuk and Conroy 1987; Mason and Macdonald 1991). However, their effectiveness in detecting distribution changes at a national level has been demonstrated by the follow-up surveys in Britain (Strachan *et al.* 1990), while in Ireland O'Sullivan *et al.* (1993) concluded that the standard otter survey technique was very efficient in most cases.

The fact that, in Ireland, otters live within city limits probably reflects the population pressures extant in other more optimal areas of habitat (Mason and O'Sullivan 1992). So the increase in the number of sites containing spraints in the Dublin area may be seen as a positive indicator of the wider picture of otter distribution in Ireland.

With regard to the national situation, otters are widespread throughout Ireland and appear to be thriving. However, recent studies by Mason and O'Sullivan (1992) and O'Sullivan *et al.* (1993) have shown that a small number of the otters they examined were carrying potentially damaging burdens of organochlorine pesticides and polychlorinated-biphenyls (PCBs). Commercial fish nets, crustacean traps and road traffic are known to cause substantial otter mortality in Britain and Continental Europe (Jefferies *et al.* 1984; Madsen 1990) and, although data on the impact of accidental death on otter populations in this country are not currently available, it is probably of some significance.

With regard to water quality in Ireland, between 1981 and 1990 there was a 15% rise in the freshwater channel length classified as either slightly or moderately polluted and, in particular, there was an increase in the eutrophication of rivers and streams (Clabby *et al.* 1992). While there is no evidence from the present study that this water-quality trend has caused any change in the status of Irish otters over the last decade, future monitoring of the situation is desirable.

The otter population in Ireland has, for some years, been regarded as one of international importance (Mason and MacDonald 1986; Foster-Turley *et al.* 1990). So, in addition to continued research on Irish otters, improving the legal protection for otters and their wetland habitats is probably the most urgent conservation measure required.

ACKNOWLEDGEMENTS

I should like to express my gratitude to Avonmore Foods plc for their generous support, without which this project could not have been undertaken. Thanks are also due to the Wildlife Branch of the Office of Public Works for their financial and practical support. I am indebted to the Irish Wildlife Federation and the Ulster Wildlife Trust for their help and to the Department of Zoology, Trinity College, Dublin, particularly my supervisor, Dr Julian Reynolds, and Dr John Rochford.

REFERENCES

Andrews, E. and Crawford, A.K. 1986 *Otter survey of Wales 1984–85.* London. Vincent Wildlife Trust.

Chanin, P.R.F. and Jefferies, D.J. 1978 The decline of the otter *Lutra lutra* L. in Britain: an analysis of hunting records and discussion of causes. *Biological Journal of the Linnean Society 10,* 305–28.

Chapman, P.J. and Chapman, L.L. 1982 *Otter survey of Ireland 1980–1981*. London. Vincent Wildlife Trust.

Clabby, K.J., Lucey, J., McGarrigle, M.L., Bowman, J.J., Flanagan, P.J. and Toner, P.F. 1992 *Water quality in Ireland 1987–1990. Part one: General assessment.* Dublin. Environmental Research Unit.

Conroy, J.W.H. and French, D.D. 1987 The use of spraints to monitor populations of otters. *Zoological Society of London, Symposium* 58, 247–62.

Crawford, A., Jones, A. and McNulty, J. 1979 *Otter survey of Wales 1977–78*. Lincoln. Society for the Promotion of Nature Conservation.

Foster-Turley, P., Macdonald, S.M. and Mason, C.F. 1990 *Otters: an action plan for their conservation*. Gland, Switzerland. International Union for the Conservation of Nature and Natural Resources.

Green, J. and Green, R. 1980 *Otter survey of Scotland 1977-79.* London. The Vincent Wildlife Trust.

Green, J. and Green, R. 1987 *Otter survey of Scotland 1984–85.* London. The Vincent Wildlife Trust.

Jefferies, D.J. 1986 The value of otter *Lutra lutra* surveying using spraints: an analysis of its successes and problems in Britain. *Otters, 1985,* 25–32.

Jefferies, D.J. 1989 The changing otter population of Britain 1700–1989. *Biological Journal of the Linnean Society* 38, 61–9.

Jefferies, D.J., Green, J. and Green, R. 1984 *Commercial fish and crustacean traps: a serious cause of otter* Lutra lutra *(L.) mortality in Britain and Europe.* London. Vincent Wildlife Trust.

Kruuk, H. and Conroy, J.W.H. 1987 Surveying otter *Lutra lutra* populations: a discussion of problems with spraints. *Biological Conservation* 41, 187–94.

Lenton. E.J., Chanin, P.R.F. and Jefferies, D.J. 1980 *Otter survey of England 1977–79.* London. Nature Conservancy Council.

Madsen, A.B. 1990 Otters *Lutra lutra* and traffic. *Flora og Fauna* 96 (2), 39–46.

Mason, C. F. and Macdonald, S.M. 1986 *Otters: ecology and conservation.* Cambridge University Press.

Mason , C. F. and Macdonald, S.M. 1991 Assessment of otter *Lutra lutra* survey methods using spraints. In C. Reuther and R. Rochert (eds), *Proceedings of the V. International Otter Colloquium: Habitat* 6, 167–9. Germany. Hankensbuttel.

Mason, C.F. and O'Sullivan, W.M. 1992 Organochlorine pesticide-residues and PCBs in otters *Lutra lutra* from Ireland. *Bulletin of Environmental Contamination and Toxicology* 48 (3), 387–93.

O'Connor, F.B., Barwick, D., Chanin, P.R.F., Frazer, J.F.D., Jefferies, D.J., Jenkins, D., Neal, E.G. and Sands, T.S. 1977 *Otters 1977.* London. Nature Conservancy Council.

O'Sullivan, W.M., Macdonald, S.M. and Mason, C.F. 1993 Organochlorine pesticide residues and PCBs in otter spraints from Southern Ireland. *Biology and Environment: Proceedings of the Royal Irish Academy* 93B (1), 55–7.

Strachan, R., Birks, J.D.S., Chanin, P.R.F. and Jefferies, D.J. 1990 *Otter survey of England 1984–86.* London. Nature Conservancy Council.

In: J.D. Reynolds (ed.) 1996 *The conservation of aquatic systems*, 117–133. Dublin. Royal Irish Academy.

OTTER CONSERVATION: FACTORS AFFECTING SURVIVAL, WITH PARTICULAR REFERENCE TO DRAINAGE AND POLLUTION WITHIN AN IRISH RIVER SYSTEM

W.M. O'Sullivan

Department of Zoology, University College, Cork

ABSTRACT

Declines in populations of otter (*Lutra lutra*) in Europe are generally believed to have been caused by a combination of factors, including habitat destruction, pollution, human disturbance and over-exploitation. Whilst the Irish otter population is considered of international importance, little attention has been given to threats facing it. In particular, there is a lack of quantitative data on the impacts or otherwise of such factors as drainage and pollution. This paper reviews the broad threats to otters in Europe and Ireland, presents data on the impacts of drainage and pollution on otter habitat and prey in an Irish riverine habitat (R. Awbeg, Co. Cork), and reviews the results of ongoing studies on the impacts of toxic bioaccumulative substances on the species in Irish aquatic ecosystems. A series of priorities/procedures for conservation of otters and aquatic ecosystems in Ireland is suggested.

INTRODUCTION

There have been dramatic declines in populations of otters (*Lutra lutra*) in Europe and these declines are most evident in countries that are heavily industrialised (Mason and Macdonald 1990), although involvement of other factors is also suggested (Chanin and Jefferies 1978; Mason 1989; Jefferies 1989).

The major and specific factors listed as threats to otters in Europe are given in Table 1. Habitat destruction is clearly cited most frequently as a major threat, followed by water pollution. More specifically, the form of habitat destruction is primarily caused by river and wetland drainage and to a lesser degree by urbanisation and various abstraction processes. Both

Table 1—The major and specific threats (percentage of times listed) to otters in 29 European countries/regions, ranging from Ireland to Siberia.

Major threats	%	*Specific threats/areas of conflict*	%
Habitat destruction	28	River/wetland drainage	17
Water pollution	25	Sand/gravel abstraction	3
Mortalities/illegal killings	19	Water abstraction	1
Recreation/disturbances	13	Urbanisation	5
Hydroelectric schemes	5	Organic pollution	14
Aquaculture/fisheries	5	Industrial pollution	14
Oil spillages	1	Acidification/forestry	3
American mink	1	Poisonous marine algae	1
		Aquaculture/fisheries	8
		Fyke nets/fish traps	7
		Mammal traps	5
		Hunting/killing	8
		Road traffic	9
		Angling	2
		Boating	1
		Tourism	2

(Data source: modified from Foster-Turley *et al.* (1990). The criteria used to classify threats are not necessarily mutually exclusive.)

organic and industrial pollution appear to be of equal importance relative to one another. Cumulatively, illegal killings and other causes of mortality can be quite significant, particularly those caused by aquaculture/fisheries developments and road traffic.

For an effective conservation strategy it is essential to determine which factors may be limiting existing populations of otters. Jenkins (1982) suggested that the distribution of otters within the western Cleddau catchment in Wales had been affected by losses of wetland and woodland as a result of river management, but overall little is known about the impact of varying levels of habitat alteration on otters (Mason and Macdonald 1986).

In Ireland, where otters are still widespread and common (Lunnon and Reynolds 1991; O'Sullivan 1991), drainage and river management schemes have been extensive. Since the Arterial Drainage Act (1945), an estimated 250,000 hectares of land have been influenced by drainage (Cabot 1985).

In general, drainage has been concentrated in regions with high proportions of marginal agricultural land. The present study focused on one drained river, the R. Awbeg, which is a tributary of the R. Blackwater in north Cork. The upper reaches of the Awbeg have also been some of the more polluted watercourses in County Cork since the 1970s (Toner *et al.* 1986).

Few studies in Europe have simultaneously investigated the impact of physical and chemical changes on aquatic habitat and the extinction of local otter populations (Olsson and Sandegren 1991). Persistent and bioaccumulative substances such as organochlorines, PCBs and heavy metals are considered the main toxic threats to otters (Mason and Macdonald 1986; Foster-Turley *et al.* 1990).

Chanin and Jefferies (1978) suggested that the use of aldrin and dieldrin in Irish agriculture did not cause a significant decline in the otter population as evidenced in Britain in the 1950s. Although the historical use of pesticides in Irish agriculture has been low, it is steadily increasing (Cabot 1985) and the impacts, or otherwise, on Irish aquatic ecosystems have not been adequately assessed. A broad investigation of organochlorine, PCB and heavy metal levels in otters and in otter habitat has recently been undertaken in south-west Ireland (Mason and O'Sullivan 1992; 1993a; 1993b; O'Sullivan *et al.* 1993).

This paper focuses on the impacts of habitat alteration and pollution on the otter population of a relatively small river system. Detailed information on these aspects of aquatic ecosystem dynamics and otter conservation is clearly required.

STUDY AREAS AND METHODS

The study centred on 14.2km of the R. Awbeg, which was examined in 1986 and 1987 (Fig. 1). The upper reaches of the Awbeg flow through five bogland systems and the area is flat and relatively exposed. The underlying geology is primarily one of Carboniferous limestone, and it is the only tributary within the Blackwater catchment to hold freshwater crayfish (*Austropotamobius pallipes*). These upper reaches were initially 'arterially' drained during the 1950s and since then 'remedial' or maintenance drainage has been carried out on the system at irregular time intervals on selected reaches. The upper 3.3km (A–B in Fig. 1) of the present study area was remedially drained during summer in 1986.

Within the study area, fish present included salmonids and eel (*Anguilla anguilla*) and the smaller fish species, stickleback (*Gasterosteus aculeatus*), minnow (*Phoxinus phoxinus*) and stoneloach (*Nemacheilus barbatula*). Information on the relative status of fish populations was obtained along the study area from 20 qualitative and quantitative electrofishings at 11 sites, each 50m long, between 1986 and 1987. Stop nets were placed across the top and bottom of sites that were quantitatively fished. The size of the total population present was estimated using the depletion method described by Seber and Le Cren (1967).

Chemical data on water quality of the Awbeg were provided by Cork County Council and additional water samples from selected sites during the survey were also analysed (DO, BOD, NH_3-N, NO_2, O-PO_4). The data were assessed in relation to EC directive 78/659 EC and standards for salmonid rivers as detailed in the River Blackwater Catchment Management Plan (Cork County Council 1989); water quality status was ascribed to four classes

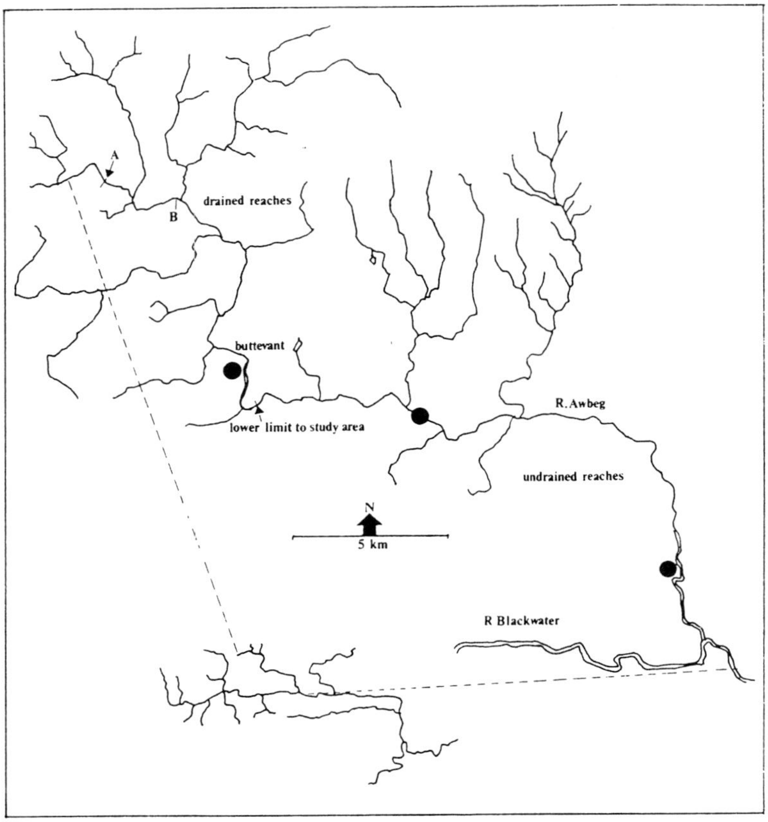

Fig. 1—Location of study area on the R. Awbeg within the Munster Blackwater Catchment. A–B = length of watercourse affected by recent drainage; solid circle = small towns villages.

ranging from heavy or gross pollution (1) to clean (4). The presence of agricultural and municipal discharges and sewage fungus was also recorded.

Both banks of the Awbeg were systematically searched for otter spraints (faeces), anal secretions and tracks. Sites at which signs were found were recorded directly onto maps of the area (scale of 1:25,000). Records were also kept of direct observations of otters.

Throughout 1987 qualitative examination of spraints in the field along the R. Awbeg was carried out. Prey items occurring in spraints were primarily categorised into the main prey groups, i.e. fish, crayfish, frogs, mammals and birds. It was possible to categorise fish prey further into individual fish families/species based on highly diagnostic head and vertebral bones; for example, opercula, dentary, maxillae and vertebrae of salmonids and eels, pharyngeal teeth of minnows, and lateral scutes of stickleback. The results are expressed for each food category in two ways—frequency occurrence (the number of occurrences of the food category recorded in all spraints) and percentage frequency (the number of occurrences of the category, expressed as a percentage of all occurrences of all items).

These procedures were used in the field to identify trends in otter diet throughout the entire 14.2km of the Awbeg study area, during September 1987.

RESULTS

Remedial drainage impacts on otter habitat and prey

During remedial maintenance, production of a more homogenous bank profile than existed previously was evident. The width of the river was increased, in places, by *c.* 30%, and the top layer of substrate (*c.* 25–30cm deep) was completely removed. In addition, all emergent macrophytes (e.g. *Schoenoplectus lacustris, Sparganium erectum, Glyceria maxima, Iris pseudacorus,* Umbelliferae) and submerged macrophytes (e.g. *Ranunculus* spp, *Callitriche* spp, *Elodea canadensis*) and occasional stands of willow/ alder were removed. Both bank margins were essentially cleared of vegetation cover, thereby removing shelter belts known to be used as the temporary resting sites by otters (personal observation).

Density estimates of fish stocks before and after remedial drainage at the Annagh study site are given in Table 2 together with comparable fish densities at the undrained site at Buttevant, 13km downstream. Stickleback clearly predominated the fish fauna at the Annagh site, both before and after remedial drainage. Densities of stickleback at the site were some 20 to 25 times higher than those at the undrained Buttevant site. Stoneloach was the second most abundant fish at the Annagh site (before drainage) followed by brown trout (*Salmo trutta*) and minnow. Salmon (*Salmo salar*) were absent. Excepting eels, the density of the fish species present showed reductions of between 77% and 100% when sampled less than two months following remedial drainage, with an overall reduction of 78% in the total fish community. When the site was again sampled, one year later, the total community density (primarily stickleback) had recovered substantially, though it was still below corresponding pre-drainage estimate. In addition, loach recovery was poor and trout, minnow and lamprey (*Lampetra fluviatilis*) were absent. In contrast, at the undrained Buttevant site, salmon were present and overall densities of salmonids were relatively high. Here the fish fauna was dominated by eels.

Otter diet over 14km of the Awbeg

A total of 715 occurrences of prey items was recorded in 590 spraints. Overall, fish formed 62.5% of prey items, of which stickleback (39.4%) were the most frequent, followed by eel (10.9%), stoneloach (5.4%), salmonids (4.6%) and minnow (2.2%). Crayfish was the second most frequently recorded prey overall (33.1%). Birds, frog (*Rana temporaria*) and dytiscid beetles comprised the remainder of the diet (4.3%).

The frequency occurrence of the main prey items in otter spraints per kilometre of watercourse is given in Fig. 2. Except for the first kilometre

Table 2—Relative fish abundance, pre- and post-remedial drainage on the R. Awbeg.

	Site A: Annagh Bridge							*Site B: Buttevant*	
	Pre-drainage July 1986		Post-drainage September 1986			Post-drainage July 1987		Undrained May 1987	
Fish species	*Density in 50m*	*Density m^{-2}*	*Density in 50m*	*% Reduction in abundance*	*Density m^{-2}*	*Density in 50m*	*Density m^{-2}*	*Density in 50m*	*Density m^{-2}*
Stickleback	426	1.162	87	77	0.217	377	0.942	18	0.045
Stoneloach	30	0.082	5	83	0.012	12	0.030	51	0.127
Brown trout	10	0.027	0	100	0.000	0	0.000	43	0.107
Salmon	0	0.000	0	–	0.000	0	0.000	35	0.087
Eel	7	0.019	12	–	0.030	6	0.015	128	0.320
Minnow	8	0.022	0	100	0.000	0	0.000	66	0.165
Lamprey	4	0.010	0	100	0.000	0	0.000	4	0.010
Total community	485	1.322	104	78	0.259	395	0.987	345	0.861

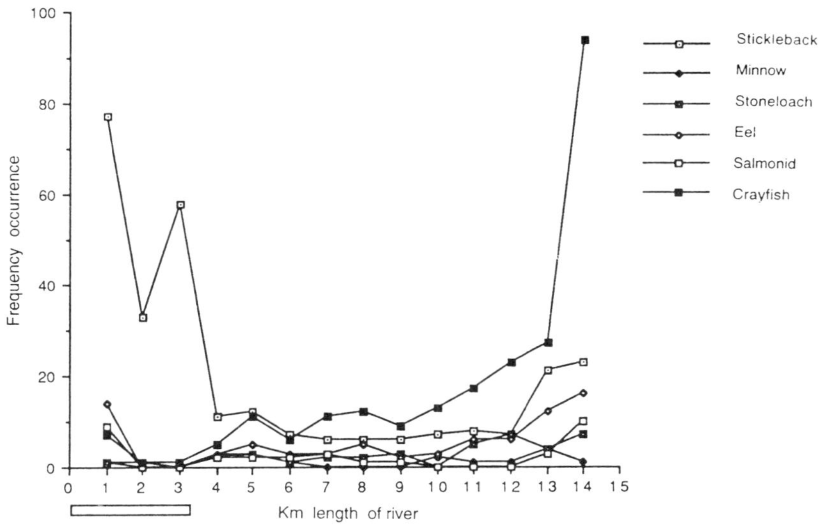

Fig. 2—The frequency occurrence of main prey items in otter spraints (n=590) per km over 14km of the R. Awbeg in September 1987.

▭ indicates length affected by recent drainage

length at the upper end of the study area and the last two kilometre lengths, recorded occurrence of salmonids and eels in otter spraints was below 10%. In addition, salmonids were absent from sprains in five kilometre lengths and eels from two kilometre lengths. Minnow occurrence was rare and loach occurrence more frequent, particularly in the lower 4km. The majority of bird remains identified were feathers, primarily black, and these are likely to have originated from Rallidae, either coot (*Fulcia atra*) or moorhen (*Gallinula chlorops*), both species having been regularly observed along the river margins.

Whilst stickleback was taken over the full 14km studied, the species was clearly most frequently predated over the upper 3km, where 59% of all stickleback occurrences were recorded. To illustrate this, both the percentage frequency of stickleback and the overall number of prey taxa recorded in otter spraints per kilometre of watercourse are given in Fig. 3. Over the first 3km, the percentage frequency of stickleback in otter spraints increased from 60% to 98%, dropping to 41% in the fourth kilometre and thereafter had a mean of 24.3% occurrence over the remaining 10km. The lowest number of prey taxa taken by otters was recorded in two contiguous kilometres coincident with the maximum percentage frequency of stickleback in the otter diet and recent drainage. From the fourth kilometre on, the number of prey taxa recorded in the diet fluctuated less widely.

In sharp contrast, lowest frequency occurrences of crayfish in the otter diet were recorded along the first 3km and thereafter increased steadily in a downstream direction, to the extent that this crustacean comprised 61% of occurrences of prey items recorded in the last kilometre. There was a highly significant positive correlation between frequency occurrence of crayfish (y) in otter spraints per km (x) in a downstream direction which can

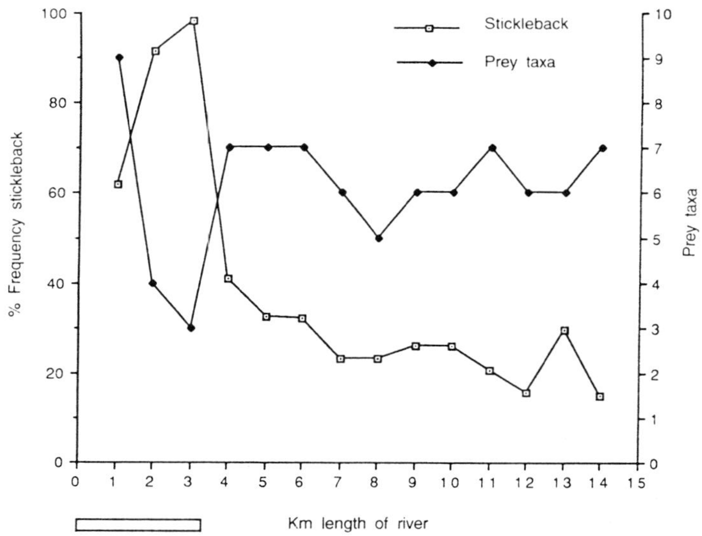

Fig. 3—Percentage frequency occurrence of stickleback and frequency occurrence of prey taxa recorded in otter spraints (n=590) per km over 14km of the R. Awbeg.

▭ indicates length affected by recent drainage

be described by the equation $y = 1.5413 \times 10^{(0.10399x)}$ ($r^2 = 0.70$, df = 12, $P < 0.002$).

Direct observations of otters after remedial drainage in summer/ autumn

Observations of a family group (adult female with young cubs) at Buttevant at a sycamore holt in May and a further family group in the middle reaches at an elm holt in July confirm otter breeding in the area. These reaches were not affected by the previous year's remedial drainage. There were three direct observations of individual otters within the study area in September. A juvenile was observed in the upper reaches in the most recently drained habitat; a dead adult female (which had been shot) was found along a well-wooded stretch in the middle reaches; and a large adult (presumably male) was observed at the lower end of the study area at Buttevant.

Pollution of the Awbeg

Overall, organic pollution was classed as heavy/gross in a total of 1.5km (10.6%) of the watercourse, moderate in 2.5km (17.6%) and low in 9.4km (66.2%); and only 0.8km (5.6%) was clean. Point discharge sources of agricultural and municipal effluent were identified at 20 and 14 locations respectively, and sewage fungus observed in a total of 2.6km of watercourse.

BOD levels of >200 mg l^{-1} were recorded at the confluence of a small tributary with the Awbeg at the upper end to the study area. The pollutant source was a combination of 'soiled water' and silage effluent. Downstream, in the main channel, BOD levels of up to 30 mg l^{-1} were recorded, a value some ten times above that expected for unpolluted river water (Cabot 1985). Sewage fungus was recorded for up to 1.7 km downstream of the effluent discharge. In addition, an even stronger effluent source (BOD value >2,000 mg l^{-1}) was identified a further 1.3 km downstream. As this stretch of river had recently been remedially drained, the water table had also been lowered and the assimilative capacity of the river at this point was further reduced.

Pollution and fish-kills within the Blackwater catchment

In County Cork, agricultural pollution is quite a problem and represents one of the major pollution threats to aquatic habitats. Preliminary surveys of 1100 holdings in 1988 showed that unlicensed run-off from farms in the county derive primarily from three sources (Table 3).

Table 3—Present/potential pollution sources from agricultural holdings in County Cork.

		BOD range (mg l^{-1})
Soiled water from farm yards	80%	1,500–2,000
Slurry	10%	17,000–25,000
Silage effluent	10%	*c.* 65,000

(Data source: Cork County Council).

Some 43% of these holdings were considered to be in the medium/high risk category, as potential pollution sources. Particular weather conditions are necessary for these pollution threats to be translated into pollution incidents and subsequent fish kills. A combination of low river flows and dry weather, followed by sudden spates of rain, bring an onset of fish kills. While 25% (12) of kills in Ireland between March and June 1989 were due to low water (Moriarty 1990), a further 20% were due to agricultural discharges.

An analysis of fish kills within the Blackwater catchment (Table 4) shows that, between 1980 and July 1992, 38 fish kills occurred with an estimated loss of *c.* 25,650 fish (primarily salmonids, cyprinids and eels). The length of watercourse affected in each case varied from 1km to 13km, while the majority averaged 3km. Some 92% of the kills occurred on tributaries and then mostly in the mid-north of the catchment. A breakdown of the causes of the fish kills showed that 21 (55%) were caused by agricultural discharges (primarily silage/slurry), 8 (21%) were caused (completely or in part) by industrial discharges, 5 (13%) by pesticides/herbicides, and 4 (11%) causes are unknown.

Table 4—Fish kills within the River Blackwater catchment in north Cork (1980 to July 1992).

River	*No. of kills*	*Estimated number of fish killed*
Blackwater	3	>2,500
Gradoge/Funshion	1	3,500
Gradoge	4	>4,100
Funshion	3	2,700
Awbeg	5	2,500
Awbeg Minor	2	900
Allow	3	1,750
Owentaraglen	3	1,350
Brogeen	2	1,500
Dalua	2	1,000
Deel	2	1,200
Finnow	1	1,000
Sheep	1	800
Araglen	1	350
Bride	1	00
Owenbaun	1	200
Minor tributary	1	100

(Data source: Cork County Council and personal observations.)

ENVIRONMENTAL POLLUTANTS

Three of the otters analysed for contaminants by Mason and O'Sullivan (1992) were found alive and showing signs of disorientation. One entered a house within Kanturk town in north Cork and was later found dead beside the R. Allow, a tributary of the Blackwater. Another was wandering, disorientated and apparently blind, along a main thoroughfare within Cork City and died the following day. A further animal was also found in a disorientated state before being hit and killed by a car in the city suburbs. All three otters had elevated levels of organochlorines and PCBs in liver samples (Mason and O'Sullivan 1992). Generally, otters from Cork City and suburbs had raised levels of organochlorine and PCB contaminants. For example, mean concentrations of dieldrin, DDT and PCBs were 1.62, 5.76 and 15.17 mg kg^{-1} (lipid) in rural otters, and 77.07, 69.04 and 118.16 mg kg^{-1} (lipid) in Cork City otters, respectively. Analysis of otters for heavy metal residues showed these to be generally low (Mason and O'Sullivan 1993a), although three otters had elevated levels of mercury potentially high enough to cause sub-lethal effects. One of these otters came from Cork Harbour and two from within the Blackwater catchment. Of the Blackwater animals, one derived from the rural Freemount area and another from the outskirts of Mallow.

DISCUSSION

Influences of drainage and pollution on habitat resources

Arterial drainage consists of deepening and widening river channels to accommodate existing river flow in order to reduce flooding and to facilitate field drainage (Cabot 1985). The character of the rural Irish landscape can be altered by: (i) the creation of spoil banks alongside drainage channels; (ii) loss of wetland; (iii) disturbance and loss of spawning areas for salmonids; and (iv) 'canalisation' of river courses, which reduces habitat diversity required to support a varied aquatic community. In general, drainage of rivers is capable of causing more thorough and more frequent disruption of a river system than any natural phenomenon. As a result, only species from pioneer or early successful stages can survive. River management following drainage often occurs before community diversity has recovered. Consequently the community is kept in a species-poor state (Andrews and Lewis 1984).

The studied habitats along the R. Awbeg are quite varied and in general drainage impacts are maximal on the upper reaches and minimal on the lower reaches, affording a gradual improvement in prey and cover resources along a continuum in a downstream direction.

Brown trout, stoneloach and minnow, present at one electrofished site before remedial drainage, were not in evidence there one year later. Other studies in Ireland have shown decreases in abundance of fish stocks following drainage (Toner *et al.* 1965; Kennedy *et al.* 1983). Similar management on an English river effectively reduced the standing crop of fish by 76% (Swales 1982). Such extensive habitat modifications have been carried out on various selected reaches of the upper Awbeg every two years and sometimes more frequently. It is clear that there are no 'ecological guidelines' given to workers by the relevant river management authorities, who are exempt from such by various planning laws. Consequently, for many stretches of the upper Awbeg, the dynamic equilibrium of the river is effectively unbalanced by remedial drainage and further disturbed by pollution levels with consequent influences on otter utilisation of the habitat.

In extreme cases of pollution, when dense algal and sewage fungus growth out-competes the marginal aquatic beds, in addition to the dramatic reduction in water quality there may be severe effects on numbers of invertebrates, fish and top predators. In 1985, cumulative agricultural effluent discharges into the eastern limb of the Awbeg were responsible for a relatively large fish kill (> 1,500). Elsewhere within the Blackwater catchment fish kills are quite common and cumulatively represent the loss of thousands of fish, reducing the potential prey density available to otters.

The impact of organic pollution on potential prey is likely to be more acute along small watercourses, where assimilative capacity is low. It is here also that density and abundance of alternative prey such as amphibia, mammals and birds are likely to be lower than in habitats fringing larger watercourses. The potential damage agricultural pollution can have on prey resources has been recognised elsewhere (Mason 1989) and it is clear that this factor can be significant in rural areas, away from dense human population.

Several Irish otters in the studies of environmental pollutants had accumulated high pollution burdens, which could have potentially sub-lethal effects and impair reproduction. Heretofore, such substances were not considered a threat to the species in Ireland and it is clear that an ongoing monitoring programme in this area is also needed.

Otter breeding, feeding habits, dietary requirements and the role of prey species

Field observations suggest that 6–8 otters (four of which were cubs) were using the 14.2km of watercourse over the summer and autumn periods of 1987 and that at least three otters were using the area in September. The observations confirm breeding in two kilometre lengths that were both relatively well wooded and unaffected by recent remedial drainage. At Buttevant, otter breeding has been recorded for three successive years (1985–7); evidence of breeding upstream was less frequent.

In undrained reaches of the Awbeg, predation on crayfish is clearly highest from May to July, when this crustacean can comprise up to 80% of the otter diet (personal observation). Similar patterns in otter predation on crayfish have been reported by McFadden and Fairley (1984) and Kyne *et al.* (1989). Crayfish are a shared food resource by otters and closely related mink (Kyne *et al.* 1989). Population studies on mink (*Mustela vison*) in the Irish midlands by Smal (1991) suggest that habitat structure and water quality influenced crayfish abundance, which in turn was a primary factor in determining mink density and residence time.

On the Awbeg it is clear that, in the upper reaches in particular, both resting sites and food resources are at a minimum. Here, the available prey base is smallest, both in species number and abundance. In general, stickleback densities are highest in summer (personal observation). At one site in July 1987 stickleback had a mean wet weight of 0.75g. Otters require an estimated 1.0–1.5kg of food each day (Erlinge 1968; Wayre 1979; Mason and Macdonald 1986). Assuming a minimum intake of 1kg of food per day, 80% of which is fish (Mason and Macdonald 1986), then an individual otter primarily feeding on stickleback on the recently drained reaches of the Awbeg would theoretically require more than 1060 stickleback per day to satisfy its dietary requirements. In marked contrast, in undrained reaches at Buttevant in September mean weight of brown trout was 124.6g. Using the same assumption, only 6.4 trout of this size would provide 80% of the estimated daily requirements of an otter. Such a predation level on stickleback would probably deplete the standing fish stocks in a relatively short period of time in the absence of high fish production levels. Moreover, other piscivores such as heron (*Andrea cinerea*) and kingfisher (*Alcedo atthis*) also feed in the same area and the latter species has also been observed feeding on stickleback (personal observation).

In general, stickleback densities are lowest in winter (Wootton 1984), which on an unmanaged river would probably have minimal impact on otter feeding behaviour or presence, as higher prey diversity and availability of alternative food would normally be expected. Of 17 otter dietary studies in Europe, reviewed by Mason and Macdonald (1986), only three studies conducted in two areas—the Somerset levels and Norfolk broads in England—recorded significant quantities of stickleback. Indeed, Kyne *et al.* (1989) expressed surprise at not recording stickleback as otter prey in the Irish midlands, where mink were clearly feeding on the species. Stickleback have been recorded in otter diets elsewhere in the Blackwater catchment (O'Sullivan 1994) and in other Irish habitats (Fairley 1972; Fairley and Wilson 1972) but in small quantities. Kyne *et al.* (1989) suggested that with other larger fish species readily available the otter, a more specialised aquatic predator than mink, largely ignores these tiny fish.

It would seem that low densities of stickleback over winter and early spring and the absence of alternative prey in sufficient quantities are factors that are likely to limit otter utilisation of the most heavily managed upper reaches of the Awbeg. There is some evidence to suggest that this is the case. Over the winter months of 1987 a considerable reduction in otter sprainting activity was noted, which was not as pronounced elsewhere in the Awbeg catchment. Also, a case of cub abandonment was observed in the area. The cub was observed over a period of three days and, despite veterinary assistance, died shortly afterwards. In another instance, in winter an adult otter which was moribund and apparently starving, was observed along the middle reaches of the study area.

CONSERVATION PRIORITIES

Much of the practical conservation carried out for the otter has been a last-ditch attempt to save the species from local extinction (Myers 1979). Essential resources for otter conservation include an adequate food supply, sleeping and breeding sites and cover for escape (Macdonald and Mason 1990). Elsewhere in Europe, as in Ireland, land drainage and river improvement schemes are still carried out in an insensitive manner with little, if any, consideration of the effects on local otter populations. Within the EC, many of these land drainage schemes are funded by the Community under the Common Agricultural Policy (CAP) and the Integrated Mediterranean Project (IMP). Reforms of the CAP may lessen the pressures somewhat on remaining otter habitats and the proposed new habitats directive may also prove effective. However, there is a tendency for resources to be concentrated on those otter populations that are fragmented and dwindling (Mason and Macdonald 1990; Prauser and Röchert 1991). The overall priorities of the IUCN/SSC Otter Specialist Group for Conservation of the Otter in Europe (Foster-Turley *et al.* 1990) are:

1. to protect those key wetlands holding substantial otter populations,
2. to press for sensitive management of river biotopes,
3. to insist on tighter controls on pollution, with particular emphasis on organochlorine contamination,

4. to collect comparative data on otter ecology and conservation in contrasting regions of the continent,
5. to emphasise the international importance of otter populations in Greece, Ireland, Portugal and Spain (countries in which the species has received little attention), and to call for more resources for the conservation of these otter populations before they suffer the same fate as those in much of the rest of Europe, and
6. to educate policy-makers and the public about the significance of wetland ecosystems.

Prauser and Röchert (1991) stressed the need for an international approach to otter research, management and conservation, suggesting that there was little benefit in concentrating protection effort on dwindling populations in Germany or parts of the UK, if the last strongholds, say in Ireland and Eastern Europe, are lost as a result of national and international policies.

The European otter is on Appendix I of CITES and is considered by the IUCN to be a vulnerable species. Internationally, the Bern and Ramsar conventions afford a certain measure of protection to otter habitat though it is not clear that the species has benefited greatly (Macdonald and Mason 1990). Under the Habitats Directive, Ireland will be required to protect threatened natural habitats, including the habitat of the otter. However, it is not yet clear how government agencies will implement the directive. Protection of areas of scientific interest (ASI) or national heritage areas will only cover a minor proportion of Irish freshwater and marine habitats. Individual reserves or protected areas will seldom be large enough to support self-sustaining otter populations (Macdonald and Mason 1990) and ideally such areas should be linked by 'corridors' of suitable habitat to allow for free movement of otters. Currently, however, it is clear that otter habitats in Ireland, especially river systems, are subject to a host of disturbances and maintenance schemes, many of which are exempt under planning laws. Indeed, such developments give little attention to the requirements of wildlife inhabiting river corridors. Both a legal framework and ecological 'guidelines' are obviously required to redress this imbalance and the following recommended baseline from which these can be developed is suggested.

1. Make public all proposed schemes.
2. Investigate alternatives.
3. Carry out cost benefit analysis.
4. Carry out Environmental Impact Assessment (EIA).
5. Incorporate biotechnological techniques at the design stage, to minimise environmental impact.
6. Conserve and, where necessary, reinstate habitat diversity/requirements of flora and fauna within wetland/river corridors, with particular reference to the maintenance of: river flows and depth; mature deciduous trees (such as oak, ash, sycamore, beech, horse chestnut); scrub and copses; dense herbaceous vegetation; macrophytes; macroinvertebrates;

Crustacea (freshwater crayfish); fish; amphibians (frog, newt, toad); mammals (particularly otter and bats); birds (e.g. sand martin, kingfisher, coot, moorhen, water rail, mute swan).

ACKNOWLEDGEMENTS

I wish to thank all who assisted with this study, especially the staff of the National Parks and Wildlife Service, the Environment Section of Cork County Council, Dr Chris Mason, University of Essex, and RPS (Cairns) Ltd, Cork. I also acknowledge the assistance of Dr Paul Giller, University College, Cork, the late Dr Pat Byrne and the late Mr Pat Behan.

REFERENCES

Andrews, J. and Lewis, G. 1984 River ecology and wildlife conservation. In G. Lewis and G. Williams (eds) *Rivers and wildlife handbook—a guide to practices which further the conservation of wildlife on rivers*, 11–18. Lincoln. RSPB, RSNC.

Cabot, D. 1985 *The state of the environment*. Dublin. An Foras Forbartha.

Chanin, P.R.F. and Jefferies, D.J. 1978 The decline of the otter (*Lutra lutra*) in Britain: an analysis of hunting records and discussion of causes. *Biological Journal of the Linnean Society* 10, 305–28.

Cork County Council 1989 *Draft water quality management plan for the Munster Blackwater catchment*. Council publication.

Council of the European Communities 1978 Council directive of 18 July 1978 on the quality of freshwater needing protection or improvement in order to support fish life (78/659/EC). *Official Journal of the European Communities*, no. L222/1.

Erlinge, S. 1968 Territoriality of the otter (*Lutra lutra* L.). *Oikos*, 19, 81–98.

Fairley, J.S. 1972 Food of otters from Co. Galway, Ireland and notes on other aspects of their biology. *Journal of Zoology, London* 166, 469–74.

Fairley, J.S. and Wilson, S.C. 1972 Autumn food of otters on the Agivey River, Co. Londonderry, Northern Ireland. *Journal of Zoology, London* 166, 468–9.

Foster-Turley, P., Macdonald, S.M. and Mason, C.F. 1990 *Otters: an action plan for their conservation*. Gland, Switzerland. International Union for the Conservation of Nature and Natural Resources/Species Survival Commission/Otter Specialist Group.

Jefferies, D.J. 1989 The changing otter population of Britain, 1700–1989. *Biological Journal of the Linnean Society* 38, 61–9.

Jenkins, A.L. 1982 *Otter distribution in the western Cleddau*. Aberystwyth. Nature Conservancy Council/Vincent Wildlife Trust.

Kennedy, G.J.A., Cragg-Hine, D., Strange, C.D. and Stewart, D.A. 1983

The effects of a land drainage scheme on the salmonid populations of the R. Damowen, Co. Tyrone. *Fisheries Management* 14, 1–16.

Kyne, M.J., Smal, C.M. and Fairley, J.S. 1989 The food of otters (*Lutra lutra*) in the Irish midlands and a comparison with that of mink (*Mustela vison*) in the same region. *Proceedings of the Royal Irish Academy* 89B, 33–46.

Lunnon, R.M. and Reynolds, J.D. 1991 Distribution of the otter (*Lutra lutra*) in Ireland, and its value as an indicator of habitat quality. In: D.W. Jeffrey and B. Madden (eds), *Bioindicators and environmental management*, 435–43. London. Academic Press.

Macdonald, S.M. and Mason, C.F. 1990 Threats. In P. Foster-Turley, S.M. Macdonald and C.F.Mason (eds), *Otters: an action plan for their conservation*.. Gland, Switzerland. IUCN/SCC Otter Specialist Group.

McFadden, Y.M.T. and Fairley, J.S. 1984 Food of otters (*Lutra lutra* L.) in an Irish limestone river system with special reference to the crayfish (*Austropotamobius pallipes*) (Lereboullet). *Journal of Life Sciences Royal Dublin Society* 5, 67–76.

Mason, C.F. 1989 Water pollution and otter distribution: a review. *Lutra* 32(2), 97–131.

Mason, C.F. and Macdonald, S.M. 1986 *Otters: ecology and conservation*. Cambridge University Press.

Mason, C.F. and Macdonald, S.M. 1990 Implementing conservation strategies. In P. Foster-Turley, S.M. Macdonald and C.F. Mason (eds), *Otters: an action plan for their conservation*, 15–17. Gland, Switzerland. IUCN/SCC Otter Specialist Group.

Mason, C.F. and O'Sullivan, W.M. 1992 Organochlorine pesticide residues and PCBs in otters (*Lutra lutra*) from Ireland. *Bulletin of Environmental Contamination and Toxicology* 48, 387–93.

Mason, C.F. and O'Sullivan, W.M. 1993a Heavy metals in the livers of otters, *Lutra lutra*, from Ireland. *Journal of Zoology, London* 231, 675–8.

Mason, C.F. and O'Sullivan, W.M. 1993b Further observations on organochlorine and PCB levels in Irish otters *Lutra lutra*. *Biology and Environment: Proceedings of the Royal Irish Academy* 93B(3), 187–8.

Moriarty, C. 1990 *Fish kills in Ireland in 1989*. Fishery leaflet 146. Dublin. Department of the Marine.

Myers, N. 1979 *The sinking ark*. Oxford. Pergamon Press.

Olsson, M. and Sandegren, F. 1991 Otter survival and toxic chemicals—implication for otter conservation programmes. In C. Reuther and R. Röchert (eds), Proceedings of the Fifth International Otter Colloquium. *Habitat* 6, 191–200.

O'Sullivan, W.M. 1991 The distribution of otters (*Lutra lutra*) within the Munster Blackwater catchment, a major Irish river system, 1988–1990. *Irish Naturalists' Journal* 23, 442–6.

O'Sullivan, W.M. 1994 Summer diet of otters on part of the River Blackwater catchment. *Irish Naturalists' Journal* 24, 349–54.

O'Sullivan, W.M., Macdonald, S.M. and Mason, C.F. 1993. Organochlorine pesticide residues and PCBs in otter spraints from southern Ireland. *Biology and Environment: Proceedings of the Royal Irish Academy* 93B(1), 55–7.

Prauser, N. and Röchert, R. 1991 Habitat destruction and otters: suggestions towards a more sophisticated approach. In C. Reuther and R. Röchert (eds), Proceedings of the Fifth International Otter Colloquium. *Habitat* 6, 201–9.

Seber, G. and Le Cren, E. 1967 Estimating population parameters from catches large relative to the population. *Journal of Animal Ecology* 36, 631–43.

Smal, C.M. 1991 Population studies on feral American mink (*Mustela vison*) in Ireland. *Journal of Zoology, London* 224, 233–49.

Swales, S. 1982 A 'before and after' study of the effects of land drainage works on fish stocks in the upper reaches of a lowland river. *Fisheries Management* 13, 105–14.

Toner, E.D., O'Riordan, A. and Twomey, E. 1965 The effects of arterial drainage works on the salmon stock of a tributary of the R. Moy. *Irish Fisheries Investigations, Series A*, 1, 36–55.

Toner, P.F., Clabby, K.J., Bowman, J.J. and McGarrigle, M.C. 1986 *Water quality in Ireland, the current position, part 1: General assessment.* Dublin. An Foras Forbartha, Water Resources Division.

Wayre, P. 1979 *The private life of the otter.* London. Batsford.

Wootton, R.J. 1984 *A functional biology of sticklebacks.* London. Croom Helm.

In: J.D. Reynolds (ed.) 1996 *The conservation of aquatic systems*, 134–135. Dublin. Royal Irish Academy.

TURBARY COMPLEX—THE FORGOTTEN FAUNA

Caroline Whelan

Department of Zoology, Trinity College, Dublin

Many of the ombrotrophic peatlands (raised bogs) on the central plain of Ireland have been exploited for fuel, chiefly in this century. One such cut-over bog, near Monasterevin, Co. Kildare, is the setting for a survey of invertebrates of cut-away pools. Formed by human activity and abandoned at different times when the peat was exhausted as a fuel source, this turbary complex of pools has become home to new invertebrate colonisers. These pools of restricted habitat diversity provide an opportunity to investigate the factors that control community structure.

The most abundant organism found was the crustacean *Asellus aquaticus*. The bivalve *Pisidium* is also the sole molluscan representative. The Diptera are well represented in turbary pools with six different families, including the Chironomidae, Culicidae, Syrphidae and Tipulidae. Coleoptera, Trichoptera and Odonata are also present. The Odonata are represented by both the Anisoptera (Libellulidae) and the Zygoptera. The Hemiptera are few in number but both *Nepa cinerea* and the family Corixidae occur.

The general species composition of turbary bog pools is richer than and rather different from that of natural pools on raised bogland, perhaps reflecting the influence of more basic ground water. Malacostracan crustaceans and molluscs are additions to the normal bogland fauna. The species composition of the sites is likely to reflect differing preferences by taxa for abiotic factors such as area, shape and depth of the pools. Regular sampling over time has increased the number of species first found, suggesting that some taxa may be using the sites at different times for parts of their life cycle.

Clustering of the data revealed that fauna of adjacent sites were generally similar. One theory under consideration is that the dispersal phase (which at these sites consists mostly of flying insects) is remaining in the locality of origin. Isolated self regulating populations may be the norm. The abundances of different taxa are modified by the interaction of species. It is thought that a number of 'core species' exist, around which the other 'satellite species' revolve. Changes in the core species may have dramatic effects on the structure of the fauna. The interactions of species may depend on:

1. predator–prey relationships
2. physical interference
3. enhancement of environment.

Further study and experimentation in these artificial peatland habitats (Whelan, in prep.) may reveal how the aquatic community of natural bog pools, currently a threatened habitat that is disappearing though drainage, is maintained.

In: J.D. Reynolds (ed.) 1996 *The conservation of aquatic systems*, 136–141. Dublin. Royal Irish Academy.

THE DISTRIBUTION, HABITAT AND CONSERVATION OF THE SMOOTH NEWT *TRITURUS VULGARIS* L. IN IRELAND

Ferdia Marnell

Department of Zoology, Trinity College, Dublin

ABSTRACT

Of the three species of amphibian found in Ireland the smooth newt has been the least studied. A national newt survey and habitat study are currently being carried out by the author. The smooth newt is found in a wide variety of habitats and has been known to breed in all types of standing water. Aquatic vegetation is important for protection and oviposition. Woodland and scrub are thought to be the favoured terrestrial habitats as they offer good cover for forays from the breeding site and plenty of refugia for winter hibernation.

INTRODUCTION

The smooth (or common) newt *Triturus vulgaris* is the only newt found in Ireland. Presumably this cold-tolerant species was able to follow the glaciers retreating northwards after the last Ice Age more closely than the other European urodeles. It is probably our only native amphibian as it seems likely that the common frog (*Rana temporaria*) and the natterjack toad (*Bufo calamita*) were both introduced.

The newt is also our most neglected amphibian. Little has been published on its distribution, habitat or ecology in Ireland. While numerous studies have been carried out in Britain, the smooth newt there is subject to competition from two other congeneric species—the great-crested newt (*Triturus cristatus*) and the palmate newt (*T. helveticus*)—which must affect its realised niche to some extent. Nonetheless, the smooth newt is common in many parts of Britain and is found in a wide variety of habitats; upland, lowland, woodland, marsh, farm, open moor and urban. It has been known to breed in all types of standing water and even in slow-moving streams (Beebee 1981; Cooke and Frazer 1976; Denton 1991; Knight 1989).

DISTRIBUTION

The smooth newt has always been under-recorded in Ireland, partly because of the elusive nature of the beast—it breeds in well-weeded ponds and is seldom encountered on land—but also through ignorance: it is regularly mistaken for a lizard, particularly during its terrestrial phase.

The first map to show the distribution of the smooth newt in Ireland was published 45 years ago (Taylor 1948). Newts were shown to be present in twelve out of 40 vice-counties and were most prevalent in the West.

In 1961 two notes in the *Irish Naturalists' Journal* (Forsyth 1961; Fenton 1961) suggested that newts were to be found in every county of Northern Ireland. An updated map in the *British Journal of Herpetology* (Taylor 1963) confirmed this, with a notable concentration of sightings around the north-east, although newts seemed to be absent from most of the rest of the country (Fig. 1). By 1973 the 10km grid system was in use and the number of

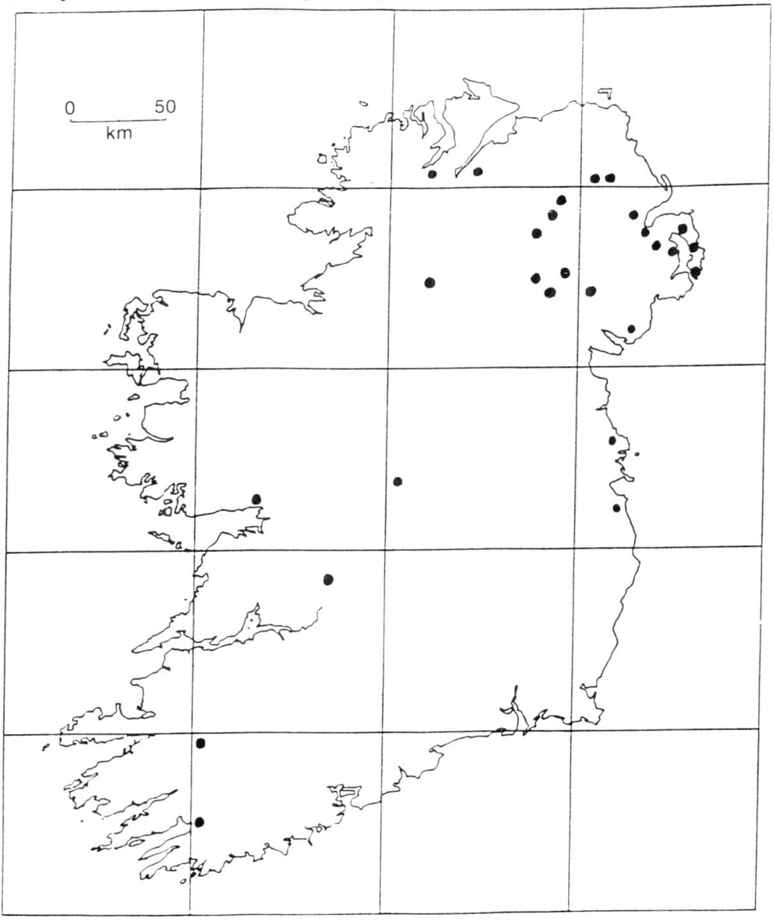

Fig. 1—The distribution of the smooth newt in Ireland from 1948 to 1963 (after Taylor 1963).

sightings had increased slightly (Arnold 1973). The animal was still thought to be far from widespread.

The first Irish-produced distribution map was published by An Foras Forbartha in 1974 (Crichton 1974). This was updated five years later (Fig. 2) with newts shown to be present in nearly one hundred 10km squares (Ní Lamhna 1979).

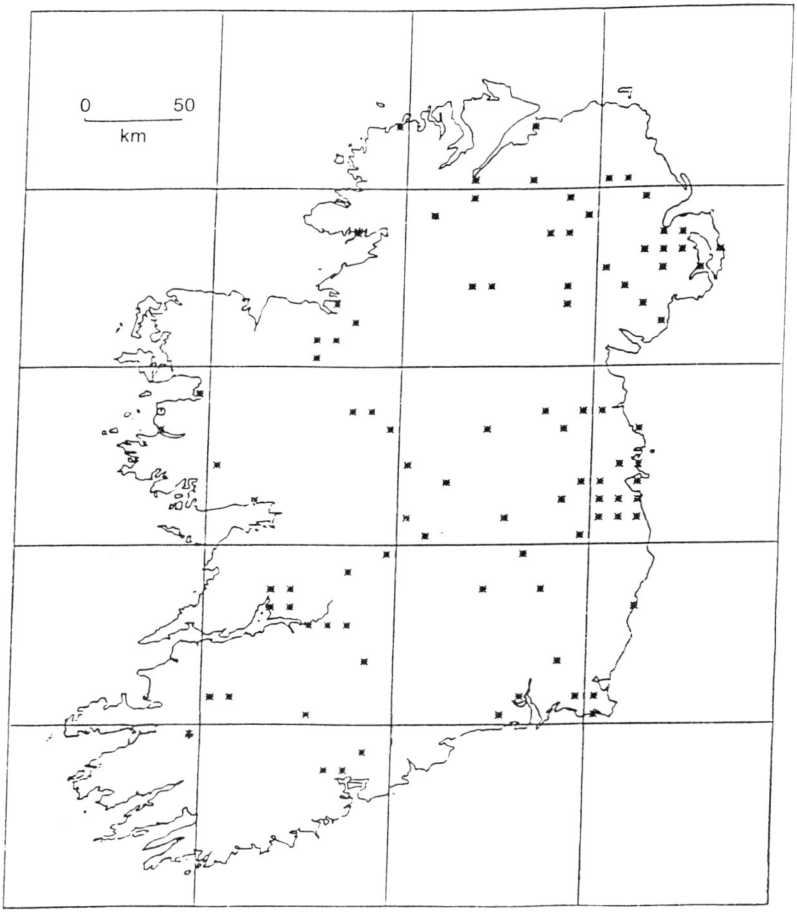

Fig. 2—The distribution of the smooth newt in Ireland from 1950 to 1978 (after Ní Lamhna 1979).

The first systematic survey of the distribution of the smooth newt in Ireland is currently underway (Marnell, in prep.) with the assistance of the Office of Public Works and the Department of Environment, N. Ireland. A stratified sampling technique, based on the National Otter Survey (Lunnon and Reynolds 1991), is being employed (see Lunnon, this volume). Initial analysis suggests that the animal is probably fairly widespread in most parts but not common anywhere.

HABITAT

The smooth newt is a highly adaptable creature; in its choice of breeding ponds it shows a wide tolerance range for pH, dissolved oxygen, conductivity and mineral concentrations. Optimum pH, however, is probably around neutral (Dolmen 1988) and below about pH 4.0 eggs will not hatch (Dunson and Connell 1982). Calcium concentrations above 15ppm and potassium levels above 1ppm may be favoured (Denton 1991).

Typical breeding ponds might be between 100m^2 and 1000m^2 and 25–100cm at maximum depth, but smooth newts have been found in water bodies 12–75,000m^2 and 10–2500cm at maximum depth (Cooke and Frazer 1976; Denton 1991; Marnell, in prep.).

Aquatic vegetation is important, not only for cover from predators and to provide cover for prey, but also for oviposition. *Glyceria fluitans* and *Callitriche* spp are particularly favoured for this function (Miaud 1990).

The terrestrial vegetation around the breeding pond has also been shown to be significant (Beebee 1981). Woodland and scrub are preferred, as they offer good cover for terrestrial forays and plenty of refuge sites for winter hibernation, but even long grass can provide enough protection and concealment.

A recent survey of N. Ireland indicated the importance of low disturbance of the breeding site and the surrounding terrestrial habitat and also a preference for fish-free ponds (Marnell 1993).

DECLINE

Amphibians are particularly sensitive bioindicators because of their permeable skin, biphasic life-cycle and pattern of embryonic development. The diversity and abundance of amphibians is declining worldwide. This is believed to be due more to habitat destruction and modification than to pollution, but acid precipitation, pesticides, changes in agricultural practices and the introduction of exotic species are also taking their toll (Wake and Morowitz 1991).

In Britain, as long ago as 1973, there was evidence that at least three of the six native amphibian species were declining significantly in numbers (Beebee 1973). The filling in of ponds, water abstraction and collection by children were suggested as probable factors in their decline.

While the distribution and abundance of the smooth newt in Ireland has not yet been established, habitat destruction and pollution, here as elsewhere, must be having a deleterious affect. Land 'improvements' in particular have probably led to the drainage of vast numbers of breeding ponds.

CONSERVATION

Garden ponds have grown to be a very important habitat for the smooth newt in suburban England and have to an extent compensated for the loss of natural breeding sites as a result of urban development (Beebee 1980).

Although most garden ponds are small, together they represent a large area of potential breeding habitat; thus amphibian-friendly garden ponds could also become a significant refuge for newts in Ireland. Such ponds should ideally have sloping sides and contain plenty of weed. Fish-free pools are highly preferable. If some of the surrounding terrestrial habitat can be left to grow wild this will also benefit amphibians.

More significantly, a conservation programme aimed at maintaining and improving rural ponds (e.g. livestock ponds, quarry pools and wetlands), particularly in areas where pond infilling and land drainage are rife, would enhance the prospects not only of our amphibians but also of the many other species of plants and animals which depend on this much neglected habitat type. However, as the smooth newt spends most of its time on land, the conservation of the aquatic system alone is no guarantee of its safety; careful management of the terrestrial habitats surrounding breeding sites is also vital.

REFERENCES

ARNOLD, H.R. (ed.) 1973 *Provisional distribution atlas for the amphibians and reptiles of the British Isles.* Abbot's Ripton, Huntingdon. Biological Records Centre, Nature Conservancy.

BEEBEE, T.J.C. 1973 Observations concerning the decline of the British amphibia. *Biological Conservation* 5, 20–4.

BEEBEE, T.J.C. 1980 More about garden ponds. *British Herpetological Society Bulletin* 1, 21–2.

BEEBEE, T.J.C. 1981 Habitats of the British amphibia 4: Agricultural lowlands and a general discussion of requirements. *Biological Conservation* 21, 127–39.

COOKE, A.S. and FRAZER, J.F. 1976 Characteristics of newt breeding sites. *Journal of Zoology* 178, 223–36.

CRICHTON, M. 1974 *Provisional distribution maps of amphibians, reptiles and mammals in Ireland.* Dublin. Folens/An Foras Forbartha.

DENTON, J.S. 1991 Newts in Cumbria. *Herpetological Journal* 1, 549–54.

DOLMEN, D. 1988 Coexistence and niche segregation in the newts *Triturus vulgaris* (L.) and *T. cristatus* (Laurenti). *Amphibia–Reptilia* 9, 365–74.

DUNSON, W.A. and CONNELL, J. 1982 Specific inhibition of hatching in amphibian embryos by low pH. *Journal of Herpetology* 16, 314–16.

FENTON, O. 1961 Distribution of the smooth newt in Ireland. *Irish Naturalists' Journal* 13, 278.

FORSYTH, M. 1961 Distribution of smooth newt, *Triturus vulgaris*, in Ireland. *Irish Naturalists' Journal* 13, 263.

KNIGHT, K.R. 1989 A survey of amphibian sites in the upland pools of mid-Wales. *British Herpetological Society Bulletin* 27, 34–6.

LUNNON, R.M. and REYNOLDS, J.D. 1991 Distribution of the otter (*Lutra lutra*) in Ireland, and its value as an indicator of habitat quality.

In D.W. Jeffrey and B. Madden (eds), *Bioindicators and environmental management.* London. Academic Press.

MARNELL, F. 1993 The distribution and habitat of the smooth newt (*Triturus vulgaris* L.) in Northern Ireland. Unpublished report to the Department of Environment (N. Ireland).

MIAUD, C. 1990 La dynamique des populations subdivisées: étude comparative chez trois amphibiens urodeles (*T. alpestris, T. helveticus* et *T. cristatus*). Unpublished Ph.D. thesis, Université Claude Bernard, Lyon.

NÍ LAMHNA, E. (ed.) (1979) *Provisional distribution atlas of amphibians, reptiles and mammals in Ireland.* Dublin. An Foras Forbartha.

SMITH, M. 1951 *British amphibians and reptiles.* London. Collins.

TAYLOR, R.H. 1948 Distribution of reptiles and amphibians in the British Isles. *British Journal of Herpetology* 1, 1–38.

TAYLOR, R.H. 1963 The distribution of amphibians and reptiles in England and Wales, Scotland and Ireland and the Channel Isles: a revised survey. *British Journal of Herpetology* 3, 95–115.

WAKE, D.B. and MOROWITZ, H.J. (eds) 1991 Declining amphibian populations—a global phenomenon? Findings and recommendations. *Alytes* 9, 33–42.

In: J.D. Reynolds (ed.) 1996 *The conservation of aquatic systems*, 142–150. Dublin. Royal Irish Academy.

IRISH PEARL MUSSELS: GOING, GOING, GONE?

Helena C. G. Chesney*

Ulster Museum, Botanic Gardens, Belfast

ABSTRACT

Ireland's rivers still have large populations of the freshwater pearl mussel *Margaritifera margaritifera*. This reflects an environment less polluted than elsewhere in Europe because of generally lower industrial, agricultural and population pressures. However, populations in a number of rivers, particularly in the north and east, have already been depleted. Such loss indicates that these aquatic systems have become unsuitable for a large range of freshwater animals, including young trout and salmon. Information on the biology, habitat preferences and requirements of *M. margaritifera* and of its ecophenotype, the hard-water '*durrovensis*' form in Ireland, supplemented by that available from populations in Germany and Russia, indicates that the large freshwater mussels, especially *M. margaritifera*, constitute a valuable element in the conservation of biodiversity of freshwater habitats. Proposals for its conservation are detailed.

INTRODUCTION

Pity the poor pearl mussels: they are not like pandas—cuddly and kind-looking—nor like the big cats—elegant and fierce—nor even like bats—menacing but vulnerable. If they were like any of these, they would by now have engendered enormous public concern over the millions killed through destruction of their habitats or the thousands killed to satisfy human greed.

However, with pearl mussels we are dealing with an animal, at first sight rather unprepossessing and even ordinary, which is at least as astonishing and intriguing as any of the above, but whose 'public profile' is almost negligible. Its pearls were well known to the Vikings and Romans and even to our grandfathers—perhaps even to our fathers—but until recently the animal, its ecology and life history were, in large part, a mystery.

The written history of pearl mussels in Ireland stretches back over one thousand years, as they were fished for their pearls and were considered a

*Formerly Ross.

precious natural resource to be harvested (or exploited). The Irish name for pearls or precious stones, *sed* or *set,* is incorporated into many placenames, such as the old name for Baltimore, Co. Cork, which was *Dun-na-sed.* Amongst the earliest recorded Irish pearls was the exquisite pearl sent in 1094 by Gilbert, the bishop of Limerick, as a gift to Anselm, the archbishop of Canterbury (Went 1947). In 1688 Sir Robert Redding wrote a most accurate and comprehensive account of pearls and pearl fishing in County Tyrone, whose rivers have been famous for pearls for centuries (Redding 1693). Smith (1750) mentions pearls 'as large as peas' being taken from mussels in Cork rivers and that the poor living nearby used the discarded shells as spoons.

The size of the populations of pearl mussels in Irish rivers can only be guessed at, but the rivers are described as having 'pavements of mussels', so tightly were they packed together (Bjork 1962). Similar descriptions from elsewhere indicate that, where conditions are optimal, the pearl mussel can live in exceptionally high densities (*ibid.*). Today, it has declined to such an extent that it is extinct, or nearly so, in several countries and is classified as 'seriously threatened' in sixteen European countries. The picture becomes alarming when data on its present whereabouts indicate a huge contraction compared with those on its former circumpolar range, extending through three continents (Wells *et al.* 1983).

Within the last decade, as the decline in its global population has become apparent, biologists and ecologists have turned their attention to the study of the animal and its habits, and since the late 1980s the results of various researches on freshwater pearl mussels in Ireland have been published, enabling us to define, more clearly, aspects of its biology and ecology that together reveal a number of unpredictable characteristics.

WHAT IS SPECIAL ABOUT PEARL MUSSELS?

Pearl mussels have evolved many unexpected and unusual traits to cope with the vagaries of the environment they inhabit. The most interesting and best-known feature about *M. margaritifera* is its natural ability, also possessed by many marine bivalves, to produce pearls, which are formed by a coating of nacre layered on to foreign matter, such as grit or small pebbles, by the mantle. It is said to be among the longest lived of any freshwater animal, with some specimens reliably aged at about 130 years. Life expectancy is important in terms of how each species copes with its environment and of its life history strategy.

An understanding of an animal's life history strategy is important when proposals for conservation of an endangered species are being considered. Perhaps the best known of several models of life history strategy are the r-K model proposed by McArthur and Wilson (1967) and the stochastic or 'bet hedging' model proposed by Mackie (1984) and Todd (1985). The characteristics of species which fit these models are given in Table 1.

The emphasis in 'r' type animals is on adult mortality, while the emphasis

Table 1—*Margaritifera margaritifera*: life history strategy compared with that predicted by r-K and stochastic (bet-hedging) models of life history strategy.

Factors	*'r' type*	*'K' type*	*Stochastic (a)*	*Stochastic (b)*	*M.m.*
Size	small	large	small	large	large
Maturity	early	late	early	late	late
Juveniles	many, small	few, large	many	few	many, small
Rep. effort	large	small	large	small	large
Longevity	short	long	short	long	long
Parent. care	no	yes	no	yes	yes
Breeding	semelpar.	iterop.	semelpar.	iterop.	iterop.

in 'K' type animals is on juvenile mortality. As *M. margaritifera* differs from these predictions in its huge reproductive effort and its production of large numbers of young, it would appear not to fit either of these strategies. The stochastic or 'bet-hedging' model depends on fluctuations in pre-reproductive survival relative to that of the adult. If there is a constant but low level of juvenile survival, the animal adopts a strategy of a short life cycle, high reproductive effort and early maturity, or semelparity. Where juvenile survival is variable, a long-lived, large species adopts a strategy of small but repeated reproductive effort or iteroparity. However, with *M. margaritifera* the small reproductive effort predicted by this theory is replaced by a large reproductive effort for a long period of time, evidenced by its longevity, the production of huge numbers of eggs, brooding of the young and high fertility throughout life (H.C.G. Chesney, unpublished).

Among the considerations that have to be taken into account to explain its response to its environment are the following.

1. It has adopted and retained a parasitic glochidial larva to overcome the problem of dispersal upstream in a lotic environment.
2. Unlike other dioecious species—where male reproductive effort is less than that of the female to ensure fertilisation in this environment—male reproductive effort is very high. It appears to be a successful strategy because up to 45% of mussels in one river were found to be gravid (Ross 1992).
3. The transfer of glochidia to fish host is a high-risk phase, subject not only to the presence of host fish of suitable (receptive) age (probably 1 or 2 year old brown trout or salmon) but also to suitable water flow.
4. Excystment also incurs high risk as the juvenile mussel must detach close to suitable substrate into which it can dig to escape predation.

The high risk at all the stages outlined above accounts for the near impossibility of predicting juvenile survival in *M. margaritifera* and the periods of very low or intermittent recruitment revealed by gaps of many years in the age classes of a population (Ross 1992). Thus neither of these models of life history strategy is useful for pearl mussels. The employment of a strategy requiring an intermediate host would seem to be a modification of the reproductive strategy typical of other large-bodied, long-lived species and to explain why the animal's life history strategy is not encompassed by either of these classic models (H.C.G. Chesney, unpublished).

However, the precarious recruitment strategy of *M. margaritifera* is compensated for by its retained ability to become self-fertile. Bauer (1988) and Ross and McCarthy (1991) report that, if populations are numerically small, hermaphrodites are produced. This contrasts with a large County Donegal population in which less than 1% was hermaphrodite (Ross 1992). By this means a low number of founder animals can establish a larger population in suitable habitats.

Another factor confirmed by recent research on *M. margaritifera* is its ability to inhabit rivers with very different pH values. In practically all published descriptions of its habitat *M. margaritifera* is described as living in soft waters, which is surprising, since heavy-shelled molluscs are normally considered to require a highly calcareous environment. The only known hard-water margaritiferid is *M. auricularia* (Spengler 1793), a close relative found in south-west Europe, which had occurred in south-east England in pre- and interglacial times, when the Rhine extended there (Preece *et al.* 1983). Access to museum collections of pearl mussels, including North American and European specimens, as well as type material of *M. durrovensis* (Phillips 1928), and *M. m.* var. *siluriana* (Bloomer 1927), allowed the soft-water requirement of *M. margaritifera.* to be examined. Stelfox (1929) and Haas (1948) considered that *M. durrovensis* was a hard-water form of *M. margaritifera*, but the uneroded shells and their very calcareous habitat indicated that a review was opportune. Additionally, the need to identify Irish endemic taxa and their current status became a priority in the light of the conservation requirements resulting from Berne Convention legislation.

Aspects of shell morphology, shell structure, anatomy and genetics of *M. margaritifera* from rivers of widely varying water chemistry were subjected to analysis. Specimens were examined from four Irish (including the Nore), two Scottish, two Welsh and one North American (Pennsylvania) rivers. The results indicated that *M. durrovensis* should be classified as an ecophenotype of *M. margaritifera* but, more importantly, that it lives in the hardest water of any known population of *M. margaritifera*. The significance of this is perhaps not obvious.

Given the geological history of Ireland and the impact of glaciation, it is unlikely that any populations of this mussel survived glaciation *in refugia*. They must, therefore, have reinvaded our rivers postglacially through the recolonisation of those rivers by the salmonid fish hosts of their glochidia, which gives a period of at most 14,000 years for the isolation and divergence of the *durrovensis* form. The average longevity—which in *M. durrovensis* is only about 40 years—is about 300 generations, a relatively short time for the isolation and subsequent evolution of a unique genotype. In fact genetic analysis showed that genetic differences between the Irish populations are not so large that one would invoke sub-species status for any. Genetic differences are almost imperceptible between the geographically close Nore and Derreen populations, but the Cork Blackwater population relates more closely to the Donegal Owencarrew population, showing that close geographic proximity does not necessarily correlate with genetic identity. There is a more distinct genetic difference between the Irish and American populations, as might be expected (Chesney *et al.* 1993). Davis *et al.* (1981), in an extensive genetic analysis of North American Unioniodea mussels including *M. margaritifera*, concluded that genetic analysis of these mussels was not definitive in solving taxonomic problems and that careful anatomical examination was required as an additional tool to aid the definition of species differences.

Though living in an unpredictable environment and having a high-risk life history strategy, including complete dependence on another animal, this species has managed to survive practically unchanged for at least two hundred million years and has also achieved an enormous geographical range (Banaresceau 1990). This must indicate that, although it may be a phylogenetic relict (Heard 1974), it has a highly evolved life history strategy, which shows supreme adaptations to life in lotic environments, and it must rank as a successful animal. However, it remains to be seen whether *M. margaritifera* can survive the alterations imposed on its environment by man.

FACTORS CONTRIBUTING TO THE DECLINE OF *M. MARGARITIFERA*

The decline in populations of *M. margaritifera* throughout its range is mainly due to eutrophication, pollution, over-fishing and modification of habitats (Wells *et al.* 1983). We have little direct evidence that these factors apply in Ireland, but we can assume that some, or all, of them have pertained where populations have declined. Fortunately two of the most serious pollution threats in Europe, namely acid rain and industrial pollution, are lower in Ireland, more because of our geographic position than for our being more environmentally aware.

A survey of pearl mussels in Northern Ireland in 1990–2 (D. Roberts, T. Mackie and H.C.G. Ross, unpublished) indicated that they have disappeared from all the northern and eastern rivers. The River Bann, where they were found in such numbers that a mussel even figured in the coat of arms of Banbridge, is totally depleted and they have almost disappeared from the Foyle catchment. This is probably due to some industrial pollution and the effects of agricultural enrichment, water course alteration and excessive siltation arising from sand and gravel extraction. This latter problem threatens the largest remaining population in Northern Ireland, in the Ballinderry River, Co. Tyrone, which will become extinct unless these activities can be curtailed.

Evidence suggests that, unless the fish hosts—in Ireland thought to be brown trout and possibly salmon—are present in sufficient numbers and of suitable age (probably under two years old), this will also have an effect on pearl mussel populations (Ross 1992). We may in this country be viewing the conservation of pearl mussels in a rather limited manner. The relationship between fish and pearl mussels may be symbiotic rather than solely parasitic. The Varzuga River in the Kola Peninsula of north-west Russia is the site of a research project on the relationship between pearl mussels and salmon, and Zyuganov and Nezlin (1991) suggest that this river produces 70% of the peninsula's salmon from only 10% of the region's total spawning area because of its immense population of M. *margaritifera* compared with other rivers in the region, which have been seriously depleted of pearl mussels by pollution and excessive water extraction. Given that each mussel filters 50l of water in 24 hours, completely purifying it, they contend that the pearl mussel and salmon together create a biocenosis resulting in a qualitatively new ecological environment, furnishing each species with optimal conditions for its existence. They are also actively pursuing the culture of mussels for their natural pearls as a commercial production of the system. The production of artificial pearls and the *in-vitro* culture of glochidia are also seen as viable projects if sufficient pump-priming funding is available.

In Ireland, as yet, there are very few definitive results from population or pollution tolerance studies which would absolutely confirm the causes of the mussel's decline. We know that the initial reduction of their population sizes started at least 100 years ago, when water quality began to deteriorate. Recent measures taken to improve river-water quality may not allow sufficient time to recolonise some rivers before the remaining populations become extinct. Evidence from studies in other countries, such as Bavaria, shows that eutrophication increases the mortality of the minute post-parasitic phase of the mussel living in the river gravels through increased growth of algae, which entrap sediment and smother them. Adult mortality probably relates to nitrate concentration in the water but, as adults are less sensitive to this than the post-parasitic juveniles, populations increasingly contain only old mussels (Bauer 1988). Some Irish populations are also over-aged with no, or very few, juvenile mussels (Ross 1984; Ross 1991).

The pearl mussel's ability to produce pearls of commercial value subjects them to another threat. In the British Isles as a whole, very few traditional pearl fishermen remain, as it is no longer commercially viable. Use of diving suits and equipment makes pearl fishing a less dangerous and uncomfortable activity. The chance of finding a valuable pearl means that hundreds, and occasionally thousands, of mussels may be killed by amateurs in one day's sport, unless its protected status is publicised.

CONSERVATION PROPOSALS

M. margaritifera is protected in Ireland under the Berne Convention of the Council of Europe, to which the Irish government is a signatory. The European Invertebrate Survey drew attention in 1992 to the biogeographic and evolutionary interest attaching not only to genetically distinct but also to infra-specific taxa of isolated or relict populations of invertebrates. The ecophenotype *M. durrovensis* of the Nore catchment is unique to Ireland, where its habitat is the highly calcareous rivers. In Germany, France and the Iberian peninsula such rivers were inhabited by *M. auricularia*, which is now practically extinct. This means that the *durrovensis* form may represent the only living margaritiferid inhabiting highly calcareous waters in Europe. Additionally, the Nore is also one of the few biotopes in Ireland where *M. margaritifera* and the freshwater crayfish (*Austropotamobius pallipes*) are found living together. These crayfish are also scheduled for conservation and habitat protection under the Berne Convention.

Conservation of biodiversity is now considered to be the most important goal in terms of global conservation. Vane-Wright *et al.* (1991) reiterate that rare or endemic fauna and flora do not exist in isolation, but usually within well defined ecosystems on which they are dependent and to which they make their own unique contributions. Therefore the total biota of an ecosystem and not just the individual taxa must be conserved in a pristine state. Biodiversity encompasses both genetic biodiversity and the adaptive diversity of biota to different biotopes. This requires urgent investigation for many animals in today's fast-changing environments. *M. margaritifera*, with its regional variations, and the endemic *M. durrovensis* form can help to further our understanding of the role of ecological and physiological adaptations which 'fit' animals for particular environments. This requires the acquisition of data, in both time and extent, from living populations on such aspects as life cycle, population dynamics, production efficiency, reproductive efficiency and genetics, for inter- and intra-specific comparison.

RESERVES

Ought the Irish government to consider setting up freshwater reserves to conserve a variety of freshwater habitats of national and international importance in areas where two or more species occur which are under threat in Ireland or Europe—for example, the River Nore watercourse? The conservation of a wide variety of freshwater biotopes is needed in order to encompass as much of the total freshwater biodiversity as possible. This would make them both more scientifically valuable and more financially viable. Reserves would require, at minimum, the acquisition of the section of a river course containing the main population of mussels, plus 30m strips on each bank, and monitoring and maintenance of the water quality upstream of each reserve. They should include a representative range of pearl mussel and crayfish biotopes.

If sufficient resources are not immediately available then the following measures to conserve pearl mussels should be implemented.

1. The Nore mussel population should be strictly protected and a public awareness campaign about freshwater mussels, similar to the successful Bat Awareness Campaign, should be undertaken.

2. A legally enforced ban on any exploitation or disturbance of pearl mussels other than for research purposes should be instituted.
3. Water quality of rivers should be improved, maintained and legally protected against all types of pollution input.
4. Water flow regimes should be maintained and river courses and river beds should be left unaltered; drainage schemes or excessive water extraction should not be permitted.
5. The stocks of brown trout and salmon under three years old should be increased and maintained.
6. River banks should be protected from cattle by fencing.
7. Trees and other suitable shading vegetation should be preserved and increased.

The remaining populations of *M. margaritifera* in Ireland are of importance in a European context (Ross 1990) and merit priority in conservation terms. In 1978, funded by the World Wide Fund for Nature and the Zoological Museum of Helsinki University, a Margaritifera Group was set up in Finland to study the distribution, ecology, morphology and conservation of this species. It would be encouraging to think that a similar group would be set up in Ireland in the near future.

REFERENCES

BANARESCU, P. 1990 *Zoogeography of freshwaters*. 1, 303–20. Wiesbaden. Aula.

BAUER, G. 1988 Threats to the freshwater pearl mussel *Margaritifera margaritifera* in central Europe. *Biological Conservation* 45, 239–53.

BAUER, G. 1991 Plasticity in life history traits of the freshwater pearl mussel: consequences for the danger of extinction and conservation measures. In A. Seitz and V. Loeschke (eds), *Species conservation—a population–biological approach*. Basel. Birkhauser.

BJÖRK, S. 1962 Investigations on *Margaritifera margaritifera* and *Unio crassus*. *Acta Limnologica* 4, 1–10.

CHESNEY, H.C.G., OLIVER, P.G. and DAVIS, G.M. 1993 *Margaritifera durrovensis* Phillips 1928: taxonomic status, ecology and conservation. *Journal of Conchology, London* 34, 267–9.

DAVIS, G.M., HEARD, W.H., FULLER, S.L.H. and HESTERMAN, C. 1981 Molecular genetics and speciation in *Elliptio* and its relationships to other taxa of North American Unionidae (Bivalvia). *Biological Journal of the Linnean Society* 15, 131–50.

HAAS, F. 1948 On *Margaritifera durrovensis* Phillips and its affinities. *Journal of Conchology* 23, 6–8.

HEARD, W. 1974 Anatomical systematics of freshwater mussels. *Malacological Review* 7, 41–2.

McARTHUR, R.H. and WILSON, E.O. 1967 *The theory of island biogeography.* Princeton University Press.

MACKIE, G.L. 1984 Bivalves. In K. Wilbur (ed.) *The Mollusca* 7, 351–418. London. Academic Press.

PHILLIPS, R.A. 1928 On *Margaritifera durrovensis,* a new species of pearl mussel from Ireland. *Proceedings of the Malacological Society of London* 18, 69–74.

PREECE, R.C., BURLEIGH, R., KERNEY, M.P. and JARZEMBOWSKI, E.A. 1983 Radiocarbon age determinations of fossil *Margaritifera auricularia* (Spengler) from the River Thames in West London. *Journal of the Archaeological Society* 10, 249–57.

REDDING, R. 1693 A letter from Sir R. Redding, late fellow of the R.S., concerning pearl fishing in the north of Ireland. *Philosophical Transactions of the Royal Society* 17, 659–64.

ROSS, E. 1984 Studies on the biology of freshwater mussels (Lamellibranchia: Unionacea) in Ireland. Unpublished MSc thesis, University College, Galway.

ROSS, E. and Mc CARTHY, T.K. 1991 The biology and conservation of the freshwater pearl mussel (*Margaritifera margaritifera*) in Ireland. *Proceedings of the 1st Irish Environmental Researcher's Colloquium, Sligo, 1991*. Galway. University College Galway.

ROSS, H.C.G. 1990 The potential role of Irish populations in the conservation of the freshwater pearl mussel (*M. margaritifera*) in Europe. *Environmental Encounters* 10, 55–8.

ROSS, H.C.G. 1992 The reproductive biology of the freshwater pearl mussel *Margaritifera margaritifera* (L.) in Co. Donegal. *Irish Naturalists' Journal* 24, 43–50.

SMITH, C. 1750 *The antient and present state of the County and City of Cork*, 355. Dublin.

STELFOX, A.W. 1929 Land and freshwater Mollusca. *Proceedings of the Royal Irish Academy* 39B, 5–10.

TODD, C.D. 1985 Reproductive strategies of north temperate rocky shore invertebrates. In P.G. Moore and R. Seed (eds), *The ecology of rocky coasts*. London. Hodder and Stoughton.

VANE-WRIGHT, R.I., HUMPHREYS, C.J. and WILLIAMS, P.H. 1991 What to protect? Systematics and the agony of choice. *Biological Conservation* 55, 235–54.

WELLS, S.M., PYLE, R.M. and COLLINS, N.M. 1983 *The IUCN Invertebrate Red Data Book,* 147–56. Gland. IUCN.

WENT, A.E.J. 1947 Notes on Irish pearls. *Irish Naturalists' Journal 9*, 41–5.

ZYUGANOV, V. and NEZLIN, L.P. 1989 Evolutionary aspects of symbiosis of pearl mussels and salmonid fishes. *Moscow Kandalaksha USSR* 3, 373–6.

In: J.D. Reynolds (ed.) 1996 *The conservation of aquatic systems*, 151–155. Dublin. Royal Irish Academy.

CONSERVATION STRATEGIES FOR THE IRISH FRESHWATER CRAYFISH

Julian D. Reynolds and Milton A. Matthews*

Department of Zoology, Trinity College, Dublin

ABSTRACT

The freshwater crayfish *Austropotamobius pallipes* Lereboullet is widespread in lime-rich waters in Ireland, while elsewhere in its range it is severely threatened. Current conservation strategies include prohibitions on capture and controls on introductions of alien crayfish which may harbour disease. In the past decade several Irish midlands populations have been lost, apparently owing to the parasitic fungus *Aphanomyces astaci* Shikora. This fungus, normally carried by American crayfish, appears to have gained access to Ireland with fishing gear, and it is to be expected that other outbreaks may follow. Depending on the value placed on this species by Irish authorities, other conservation strategies might include education (few are aware that crayfish are present), limited fishing for consumption to raise their perceived value, and stricter controls on sports fishing gear.

INTRODUCTION

The white-clawed crayfish *Austropotamobius pallipes* (Lereboullet) occurs from Britain and Ireland through France to Iberia, Italy and Yugoslavia. It is a slow-growing species, reaching 100mm total body length in around ten years (Laurent 1988).

The Irish distribution of *A. pallipes* follows the Carboniferous limestone, with populations in most parts of Ireland except the acid mountain fringe and the north-east (Reynolds 1982; Lucey and McGarrigle 1987; Laurent 1988). While it appears to be native to Ireland (Reynolds 1982) or at least a long-standing resident, it is little known to most country people, and does not appear to figure in Irish folklore (S. Reynolds, pers. comm.).

Crayfish occur in streams, ponds and lakes, where they may control growth of aquatic vegetation and have an impact on benthic invertebrate

*Current address: The Salmon Research Agency of Ireland, Farran Laboratory, Furnace, Newport, Co. Mayo.

communities (Matthews and Reynolds 1992; Matthews *et al.* 1993). They are sometimes an important food for fish (Moriarty 1963) or otters (McFadden and Fairley 1984). In recent decades there has been a dramatic decline in the abundance of European crayfish, through a combination of overfishing, pollution and habitat disturbance, but especially through the spread of the parasitic crayfish plague fungus *Aphanomyces astaci* Schikora. The white-clawed crayfish has potential value for harvest and for aquaculture (Matthews 1992; Reynolds 1979; 1988b; 1989a; 1990), but it has declined greatly in Europe, is now listed in the Berne Convention Invertebrates project (Reynolds 1989b) and is protected in Ireland.

Until recently Ireland remained the last European country free of the crayfish plague fungus (O'Keeffe and Reynolds 1983). In 1987 a trapping survey indicated recent crayfish losses from a number of Irish lakes (Reynolds 1988a), and crayfish plague was diagnosed on Lough Lene, Co. Westmeath (Grid Ref. N 51 69) (Matthews and Reynolds 1990). It is believed that the plague fungus may have been imported accidentally with foreign fishermen and their gear. The sequence of crayfish stock losses appears to be: 1985—White Lake; 1986—Loughs Bane, Glore, Owel; 1987—Lough Lene (Map 1). Other recent crayfish losses may be due to pollution (Reynolds 1988a; Matthews 1992). There is no current evidence of crayfish plague in Irish waters, suggesting that the disease may have burnt itself out (Matthews 1992). Exploitation has not been identified as a cause of crayfish decline in Ireland.

OPTIONS FOR CONSERVING CRAYFISH STOCKS

The principal threat to Irish crayfish appears to be the reintroduction of plague with resistant living crayfish species, and its perennial establishment. A number of possible approaches to protection of our freshwater crayfish stocks have been adopted or are suggested, as follows.

Quarantine controls on alien crayfish

These would protect against introduction of disease. They have been operated with success in Ireland for many years (Unestam 1975; Gibson 1979; Reynolds 1979). In Britain, where several alien crayfish species now occur, Holdich and Reeves (1991) have recommended 'no-go' areas in which transmission and culture of foreign plague-carrying crayfish species would be prohibited. This would be difficult to implement in Ireland, where the midlands catchments are often ill-defined.

Stricter controls on movement of anglers and/or disinfection of gear

Norway has successfully implemented such a programme for sports fishermen (J. Skurdal, pers. comm.), but this may not be acceptable to the Irish or visiting angler.

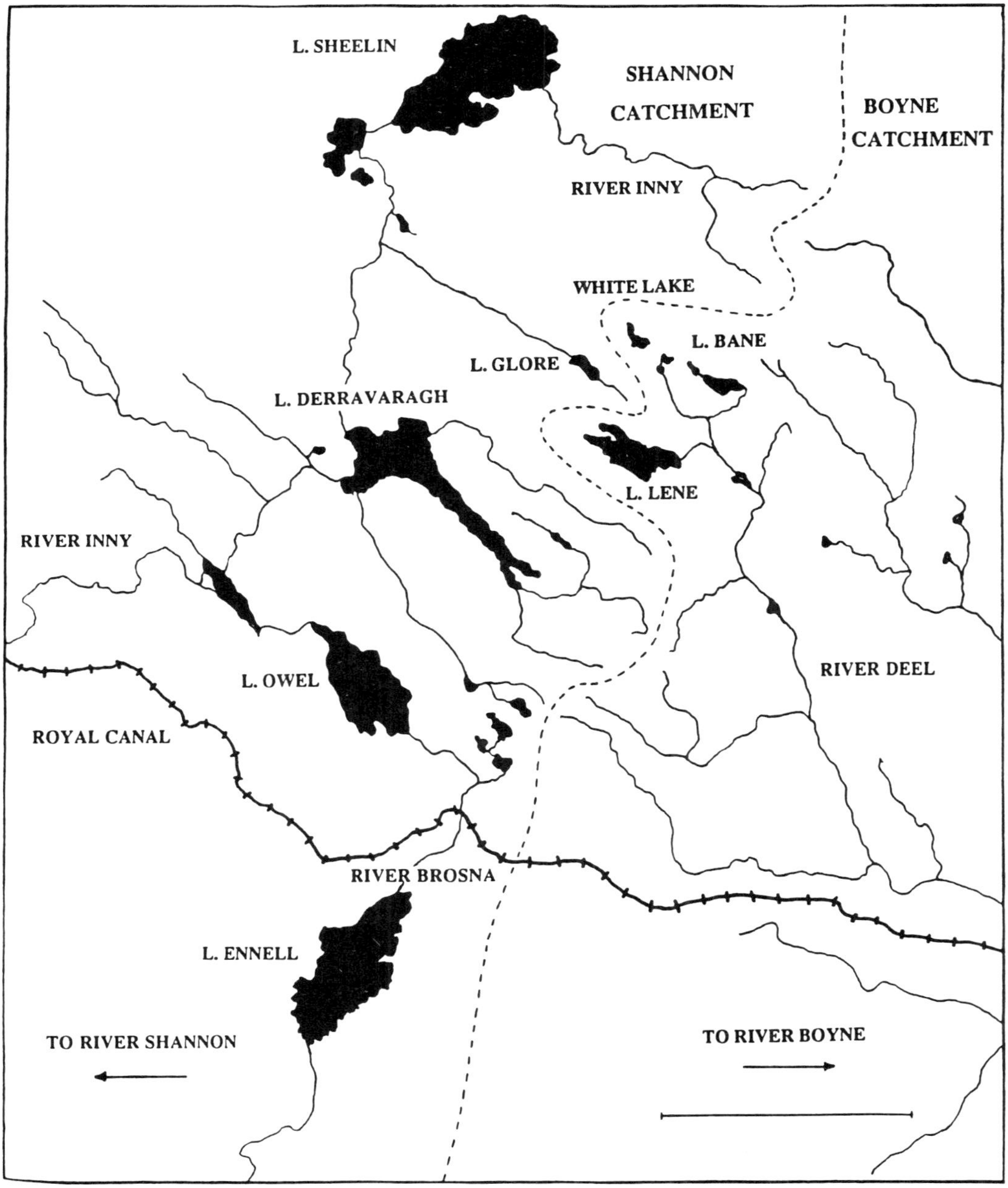

Map 1—Irish midlands lakes in the Shannon and Boyne catchments. Those which have recently lost resident crayfish populations owing to crayfish plague or other causes are named (from Matthews 1992).

Legislation against exploitations

Apart from following European legislation under the Berne Convention, there seems no rational reason for a prohibition on crayfish capture in Ireland. Indeed, permits for selective harvesting may make the public more aware of the existence of this long-lived, large invertebrate, and hence increase its perceived value.

Reintroductions

With the apparent elimination of plague, large-scale reintroductions of crayfish are feasible and, in view of their ecological role, may be desirable to maintain benthic community structure in certain lakes (Matthews 1992; Matthews *et al.* 1993). While phenotypic differences have been noted between Irish crayfish stocks (Reynolds 1989c), individuals of a stock spreading from a Wicklow stream into Blessington reservoir shows comparable changes in growth and size (Matthews 1992).

Education

Occasional accounts of the crayfish in Ireland have appeared in the press and in popular journals. More emphasis is needed on the intrinsic ecological and potential economic value of native stocks.

ACKNOWLEDGEMENTS

We acknowledge with thanks the financial and practical assistance of Salmara Ltd, the Electricity Supply Board, the Department of the Marine, and the Department of Zoology, Trinity College, Dublin.

REFERENCES

GIBSON, F.A. 1979 An international code of practice and the Irish regulations relating to certain aquatic organisms. In R.P. Kernan, O.V. Mooney and A.E.J. Went (eds), *The introduction of exotic species, advantages and problems,* 69–72. Dublin. Royal Irish Academy.

HOLDICH, D.M. and REEVE, I.D. 1991 Distribution of the freshwater crayfish in the British Isles, with particular reference to crayfish plague, alien introductions and water quality. *Aquatic Conservation: Marine and Freshwater Ecosystems* 1, 139–58.

LAURENT, P.J. 1988 *Austropotamobius pallipes* and *A. torrentium,* with observations on interactions with other species in Europe. In D.M. Holdich and R.S. Lowery (eds), *Freshwater crayfish: biology, management and exploitation,* 341–64. London. Croom Helm.

LUCEY, J. and McGARRIGLE, M.L. 1987 The distribution of the freshwater crayfish in Ireland. *Irish Fisheries Investigations,* A29, 1–13.

McFADDEN, Y.M.T. and FAIRLEY, J.S. 1984 Food of otters *Lutra lutra* (L.) in an Irish limestone river system with special reference to the crayfish *Austropotamobius pallipes* (Lereboullet). *Journal of Life Sciences of the Royal Dublin Society* 5, 65–76.

MATTHEWS, M.A. 1992 Reproduction, growth and aquaculture potential of the freshwater crayfish *Austropotamobius pallipes* (Lereboullet). Unpublished PhD Thesis, Department of Zoology, University of Dublin.

MATTHEWS, M. and REYNOLDS, J.D. 1990 Laboratory investigations of the pathogenicity of *Aphanomyces astaci* for Irish freshwater crayfish. *Hydrobiologia* 203, 121–6.

MATTHEWS, M.A. and REYNOLDS, J.D. 1992 Ecological impact of crayfish plague in Ireland. *Hydrobiologia* 234, 1–6.

MATTHEWS, M.A., REYNOLDS, J.D. and KEATINGE, M.J. 1993 Macrophyte reduction and benthic community alteration by the crayfish *Austropotamobius pallipes* (Lereboullet). *Freshwater Crayfish* 9, 289–99.

MORIARTY, C. 1963 Food of the perch (*Perca fluviatilis* L.) and trout (*Salmo trutta* L.) in an Irish reservoir. *Proceedings of the Royal Irish Academy* 63B, 1–31.

O'KEEFFE, C. and REYNOLDS, J.D. 1983 The occurrence of crayfish diseases and their significance in Ireland. *Freshwater Crayfish* 5, 299–306.

REYNOLDS, J.D. 1979 The introduction of freshwater crayfish species for aquaculture in Ireland. In R.P. Kernan, O.V. Mooney and A.E.J. Went (eds), *The introduction of exotic species, advantages and problems,* 57–64. Dublin. Royal Irish Academy.

REYNOLDS, J.D. 1982 Notes on the Irish distribution of the freshwater crayfish. *Bulletin of the Irish Biogeographical Society* 6, 18–24.

REYNOLDS, J.D. 1988a Crayfish extinctions and crayfish plague in Ireland. *Biological Conservation* 45, 279–85.

REYNOLDS, J.D. 1988b Options for crayfish culture and exploitation in Ireland. *Freshwater Crayfish* 7, 327–31.

REYNOLDS, J.D. 1989a Crayfish culture and exploitation in Ireland. In J. Skurdal, K. Westman and P.I. Bergan (eds), *Crayfish culture in Europe.* (Report from the workshop on crayfish culture, 16–19 November 1987), 92–5. Trondheim. Norway.

REYNOLDS, J.D. 1989b Current status and conservation of the white-clawed crayfish, *Austropotamobius pallipes.* In *Colloquium, the Berne Convention: Invertebrates and their conservation, Strasbourg, 8–10 November 1989*, 64–6. Report T-PVS (89) 34, 2nd rev., Council of Europe.

REYNOLDS J.D. 1989c Phenotypic variability in freshwater crayfish and its implications for aquaculture. In J.C. Aldrich (ed.), *Phenotypic responses and individuality in aquatic ectotherms,* 197-201. Ashford, Ireland. Japaga.

REYNOLDS, J.D. 1990 A perspective on crayfish culture in the British Isles. In T. Cross and B. Ottway (eds), *Proceedings of the Institute of Fisheries Management 20th Annual Study Course, September 1989,* 190–99. Galway Regional Technical College.

UNESTAM, T. 1975 The dangers of introducing new crayfish species. *Freshwater Crayfish* 2, 557–61.

In: J.D. Reynolds (ed.) 1996 *The conservation of aquatic systems*, 156–160. Dublin. Royal Irish Academy.

RATIONAL FISHERIES MANAGEMENT TO SAFEGUARD THE FUTURE

David de G. Griffith

Fisheries Research Centre, Abbotstown, Dublin

ABSTRACT

The principles of the science underlying the current policies of rational fisheries management are described. Inherent problems (for management) in the nature of the resource are identified and discussed in the context of maximising the sustainable long-term yield from a fish stock as a management objective.

SCIENTIFIC BASIS OF FISHERIES MANAGEMENT

Scientific investigations at sea and in the ports acquire knowledge on two related fronts. Data from catches are collected in a way which enables us to back-calculate to determine the size of the stock in the sea. Simultaneously, egg surveys are carried out to give a cross-check on the size of the parent stock; groundfish surveys (using trawl gear) and acoustic surveys (with calibrated echosounders) amplify and calibrate the catch sample data by providing fishery-independent estimates of stock size.

As in studies of any other biological population, information about the age structure is vital, and a considerable amount of laboratory work is therefore necessary to read the ages of the fish sampled in the field.

Comparing the relative abundance of the same cohort of fish in consecutive years (the number of three-year old cod in year x compared with the number of four-year old cod in year x+1, for example) is one way of calculating the impact of fishing on the stock. It allows a coefficient of instantaneous fishing mortality to be derived, which is essential for back-calculating from the catch data to determine the age structure and the size of the resource itself—the stock of fish in the sea—and to see how this varies from year to year.

National programmes such as these, however, build up a picture of only that part of the stock available to the national fleet. While some fish stocks are virtually entirely within our jurisdiction, such as the Celtic Sea herring, most are spread over a wide geographic area for all or part of their lives, and are consequently fished by boats of several nations. In order, therefore, to formulate scientific advice on a given fish stock as a *whole* (instead of just on a component of it, which would not be scientifically valid), the fisheries scientists of all the nations exploiting the stock in question come together to pool their data and to make a joint scientific assessment.

This is done under the aegis of the International Council for the Exploration of the Sea (ICES), an intergovernmental marine research organisation founded in 1902 with headquarters in Copenhagen. ICES organises annual sessions of a number of Stock Assessment Working Groups, and distils the groups' findings into a scientific advisory report on the state of the stocks, together with a range of appropriate management measures for each. This report is submitted every year to the seventeen member governments of ICES, as well as the EC Commission, the North East Atlantic Fisheries Commission, the International Baltic Sea Fishery Commission and the North Atlantic Salmon Conservation Organisation.

For member states of the European Community, such as Ireland, the next stage is for the scientific advice to be processed by the Commission into a package of proposals for catch regulations which will be applied in the following year. These Total Allowable Catches (TACs) are based on the scientific advice received, as are the accompanying technical measures—mesh size regulations, closure of areas and/or seasons for fishing, minimum legal landing size for fish, etc. The TACs are divided into national quotas in accordance with an allocation key agreed at the time the Common Fishery Policy was brought into being in 1983.

The process is summarised in Fig. 1.

National fisheries research and monitoring programmes
↓
Stock assessments by international scientific working groups (ICES)
↓
ICES scientific advice
↓
EC Scientific and Technical Committee for Fisheries
↓
Commission proposals for TACs and technical measures
↓
Commission Working Groups (reps. of fisheries administrations)
↓
Committee of Permanent Representatives
(representatives of national foreign ministries)
↓
Council of Fisheries Ministers
↓
Agreed management package of TACs and technical measures
↓
Implementation by national authorities

Fig. 1—Summary of the process underlying EC fisheries management regulations, showing their scientific basis.

THE NATURE OF THE PROBLEM

The availability of the resource varies from year to year, not only as a result of fishing pressure but also from natural fluctuations in stock size and composition, or altered migration patterns (determined by changes in oceanographic conditions, for example), or a mixture of all three. Thus the natural system, which we are attempting first to quantify and then to regulate, already has an inherent instability, since it depends on the number of young fish which come into the fishery each year; this annual input (known as 'recruitment') can show a natural inter-annual variation of 5–10 times.

The problem, as described in Griffith (1988), is how to take this variability into account, and establish a regime which minimises its effects. The system which we are attempting to manage has three main components: stock size, catch, and fishing effort, which are related thus:

$$\text{Catch} = \text{stock} \times \text{fishing mortality}$$

$$\text{Catch} \div \text{fishing mortality} = \text{stock size}$$

$$\text{Catch} \div \text{stock size} = \text{fishing mortality},$$

where fishing mortality is proportional to fishing effort.

Notes: (a) determining any two of these factors produces a value for the third;
(b) if fishing effort is stabilised, catches will fluctuate;
(c) if catches are stabilised, fishing effort must fluctuate.

OVERFISHING

One of the principal concerns of management is to reduce (or avoid) overfishing. This term has several definitions; the two most relevant here are:

(i) *growth overfishing*—the reduction in average weight of fish in the stock (and hence in the catch) as the exploitation rate rises;

(ii) *recruitment overfishing*—reduction of the stock to such a low level that it may not be able to generate a sufficient number of recruits to sustain itself.

The principal yardstick by which to judge whether or not a stock is being overfished is the Yield-per-Recruit curve. This is derived from the growth parameters of the stock in question, together with the age of recruitment to the fishery and the relative fishing mortality on each age group. The Y/R curve describes the yield (and the biomass in the sea) obtainable from a single year-class at various levels of fishing, assuming that a steady state exists, i.e. that the input values remain constant. In this steady state situation, the yield from a single year-class throughout its life, at any given level of fishing pressure, will be the same as the yield in one year from all the year-classes present in the stock. (Running the model as yield per *recruit* prevents natural fluctuations in year-class strength from obscuring the relationship between catch weight and fishing effort.) Estimates of long-term annual yield can be derived by multiplying Y/R by average recruitment.

The Y/R curve for Donegal haddock is given in Fig. 2 as an example. The level of fishing mortality (F = 0.2, in this case on the x axis) corresponding to the peak of the curve represents that level giving the maximum sustainable yield, i.e. the maximum long-term average annual catch which, under conditions of average recruitment, can be taken from the stock under the present pattern of relative fishing mortality across the age groups in the stock. It may be seen that the current level of fishing mortality (F = 0.85) is considerably above the optimum; fishing less intensively would allow the fish to remain in the sea longer, to grow bigger and thus boost the yield. Furthermore, the lower the level of fishing mortality, the greater the stock biomass; thus catch per unit of fishing effort (catch per hour, for example) would also be greater with less fishing, as can be deduced from the biomass-per-recruit curve.

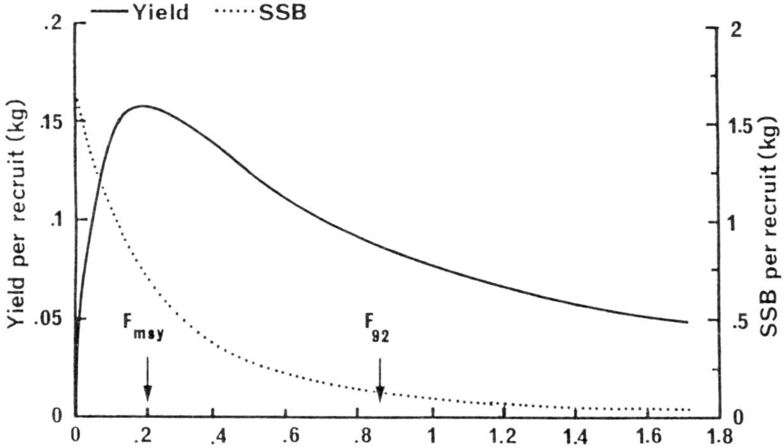

Fig. 2—Yield per recruit and biomass per recruit for Donegal haddock (SSB = spawning stock biomass, the combined weight of mature haddock in the sea).

EXPANDED RESEARCH ACTIVITIES

As discussed in Griffith (1992), the stocks on which most of our so-called 'traditional' fisheries are based are already fully exploited, or even overfished, in terms of the Y/R model. For this reason a policy of constraint on new fishing boats entering the fleet and the diversion of fishing effort on to new stocks and on to those for which our national quotas are not fully utilised has been established by the Department of the Marine. In order to quantify the sustainable yield which could be obtained from stocks of the new species (those not hitherto fished by Irish boats) additional research activities have been put in place in the Fisheries Research Centre (FRC).

Up to 1992, the staff resources available to the stock assessment programme covered only herring, mackerel, sole, haddock, two out of three plaice stocks, three out of four cod stocks and three out of four whiting stocks. Work on hake, monkfish, megrim and elasmobranch stocks had to be abandoned in 1992 because of cutbacks. The EC STRIDE Initiative, however, has enabled us to increase radically the scope of our activities to fill these shortcomings and to expand our research into a number of new areas. The most exciting of these is the acquisition, for the first time, of in-house expertise in hydroacoustics (see the reference to acoustic surveys on p.1) and of extra sea-going staff who will work with the fleet to collect valuable biological and other data on board, in addition to those being obtained in the ports. These measures will allow more effective conservation of many exploited fish stocks through up-to-date information on their biological responses to exploitation.

REFERENCES

GRIFFITH, D. de G. 1988 In *Report on the sixth dialogue meeting, October 1987,* International Council for the Exploration of the Sea, Copenhagen, Cooperative Research Report no. 158.

GRIFFITH, D. de G. 1992 Marine resources—the seas. In J. Feehan (ed.), *Environment and development in Ireland,* 401–8. Dublin. The Environmental Institute, University College Dublin.

In: J.D. Reynolds (ed.) 1996 *The conservation of aquatic systems*, 161–163. Dublin. Royal Irish Academy.

MONITORING FOR THE CONSERVATION OF EXPLOITED BENTHIC SPECIES

Oliver Tully

Department of Zoology, Trinity College, Dublin

Exploited sedentary benthic species are subject to local depletion because fishing is typically directed at areas supporting the highest biomass. This creation of depleted patches in terms of biomass, mean size and mortality may not be reflected in catch statistics for the fishery because fishing effort can be relocated to new areas, thereby maintaining total catches, and theoretically may not be apparent until the last patch is depleted. Fishing effort data are not usually collected on a spatial scale that can indicate this type of exploitation. Research surveys that sample the entire area in which the resource exists can give an early warning of an impending population crash by indicating the spatial patterns in biomass, mortality and age structure that exist and by isolating changes in the population characteristics that are brought about by exploitation. To do this the natural spatial variation in the biological characteristics of the species must be separated from that brought about by exploitation. It may be relevant, therefore, to relate the spatially variable biological parameters of the population to similar scale variation in the environment. Monitoring surveys will indicate changes in the size, number and location of areas receiving the highest level of impact from exploitation. Parameters of the population important for fisheries management, such as growth and mortality, can also be derived for each individual patch.

Identification of spatial variation in biological characteristics or patches in a sampling grid should initially proceed by indicating the scales of spatial autocorrelation between samples, i.e. the correlation between the level of similarity between samples and geographic distance separating the sampling locations, by constructing a spatial variogram (Isaaks and Srivastava 1989). The scale over which the similarity between samples is correlated with geographic distance indicates the dimensions of patches within which biological characteristics of the resource are similar and which may be receiving the same level of exploitation. Classification analysis of the sampling points on a grid, using the biological characteristics of the species under study, can be used to identify areas within which the characteristics are similar and mapping procedures can then locate these patches on the sampling grid (Tully and Hillis 1995). Furthermore, the resulting map can be superimposed on a map of the area that shows spatial patterns in environmental quality. For marine benthic species this might include a map of the sediment structure, depth or bottom water temperature contours. Similarity in the spatial patterns of both biological and environmental data is a graphical indication of a correlation between these variables, i.e. a similarity of response over the grid or sampling surface. This may be difficult to demonstrate using conventional point–point correlation analysis.

In the north-west Irish Sea the Norway lobster *Nephrops norvegicus* is abundant in an area covering approximately 750km^2 and is heavily exploited. The level of exploitation is not uniform over the entire area but is concentrated in areas where biomass is highest, where bottom currents are suitable for fishing at all times and where bottom light intensities are such that the emergence of the lobsters from their burrows is not restricted to short periods of time (Tully *et al.* 1989). The environment is also variable. Sediments vary from silt to mud or sand, depth varies from 20m to 100m and bottom water temperatures in summer are about 9°C in areas where the water column stratifies, but up to 14°C in areas where the water column is mixed. Potentially, therefore, spatial patchiness in biological characteristics could arise from spatially heterogenous exploitation or from variability in the environment. *Nephrops* length frequency distribution (LFD) data, which reflect patterns of growth, recruitment and mortality, also vary with location in the Irish Sea (Tully and Hillis 1990). Survey data from 1989 and 1990 indicated that the similarity between the length–frequency distributions was correlated with the geographic distance separating them, up to a certain distance. Male and female LFDs were spatially autocorrelated like this at distances up to 25km and 15km respectively. Two to three main patches of geographically connected stations of male and smaller patches of female LFDs, with weaker spatial contiguity, were present. Significant inter-patch differences in median length, mean weight, range in length, size at age 2 and 3, and the proportion of mature females in the population were apparent (Tully and Hillis 1990). These parameters reliably classified over 80% of sampling stations to their respective patches in a discriminant analysis.

The patch structure that was detected may have been generated by both exploitation and the environment. Fishing mortality of males is higher than that of females because ovigerous females remain in their burrows and are less catchable. (Tully *et al.* 1989). The larger scale of patches and the location of a large patch of male *Nephrops,* showing high mortality, in the area where fishing effort is highest indicate an impact of exploitation. The lower proportion of variance in the male data accounted for by the sediment compared with the female data also suggests that exploitation patterns are having a greater impact on males than females.

The existence of spatially structured variability, as discussed for *Nephrops* in the Irish Sea, indicates that various, but largely undefined, processes impose a regulatory effect on the biological parameters of benthic species. Although monitoring of future changes in the spatial patterns can be used to guide the formulation of conservation measures—it can for instance indicate change in the size of depleted patches—a better understanding of the species responses to the physical environment and to exploitation is required for proper management. These functional characteristics of the exploited system could, as discussed by Sarda (1993), be incorporated into assessment methods. This approach could in fact be extended to benthic communities in areas supporting exploited species because they are also affected indirectly by exploitation and thus provide a more comprehensive basis for conservation.

REFERENCES

ISAAKS, E.H. and SRIVASTAVA, R.M. 1989 *An introduction to applied geostatistics.* Oxford University Press.

SARDA, F. 1993 Bio-ecological aspects of the decapod crustacean fisheries in the western Mediterranean. *Aquatic Living Resources* 6, 299–305.

TULLY, O. and HILLIS, J.P. 1995 Causes and spatial scales of variability in population structure of *Nephrops norvegicus* (L.) in the Irish Sea. Fisheries Research 21, 329–47.

TULLY, O., HILLIS, J.P. and McMULLEN, D. 1989 Fitting normal curves to polymodal length frequencies to assess growth in *Nephrops.* International Council for the Exploration of the Sea, unpublished communication CM/K: 32.

In: J.D. Reynolds (ed.) 1996 *The conservation of aquatic systems*, 164–166. Dublin. Royal Irish Academy.

GLASS EEL STOCKS AND THE CONSERVATION OF EELS IN THE SHANNON BASIN, IRELAND

R.E. Donnelly and T. Walsh

Department of Zoology, Trinity College, Dublin

ABSTRACT

For the continued maintenance and development of the eel population and fishery in the Shannon catchment, migrating elvers are caught at downstream barriers and stocked into the Shannon lakes. With a decline in the numbers of elvers caught in recent years, an assessment of current stocks of glass eels (young elvers) entering the Shannon Estuary was initiated and the potential of catching glass eels for the continued stocking of the Shannon lakes evaluated.

INTRODUCTION

The conservation of the Shannon eel population is an important objective of the Fisheries Conservation Section, Hydro Group of the Electricity Supply Board, who have responsibility for all fisheries on the Shannon. To conserve the population while maintaining eel fisheries, elvers have been trapped for many years at downstream barriers, mainly at Ardnacrusha and Parteen, and stocked into the lakes of the system. In recent years there has been a decline in the numbers of elvers trapped at Ardnacrusha dam near Limerick. This trend is shown in Fig. 1, which contains data on the numbers of elvers caught between 1979 and 1992. The threshold level of elvers estimated by Moriarty (1984) to be necessary to restock the lakes has not been caught since 1982. It is believed that only a small percentage of glass eels entering an estuary may survive to the elver stage. To hope to increase the numbers of elvers which make it upstream, a greater number of glass eels must survive to this stage. A possible alternative solution is to intercept the eels further downstream, when they pass the mouth of the estuary, and transfer them direct to freshwater. As there are no records to indicate the numbers of glass eels which are entering the estuary, the first aim was to survey glass eel distribution in the Shannon Estuary.

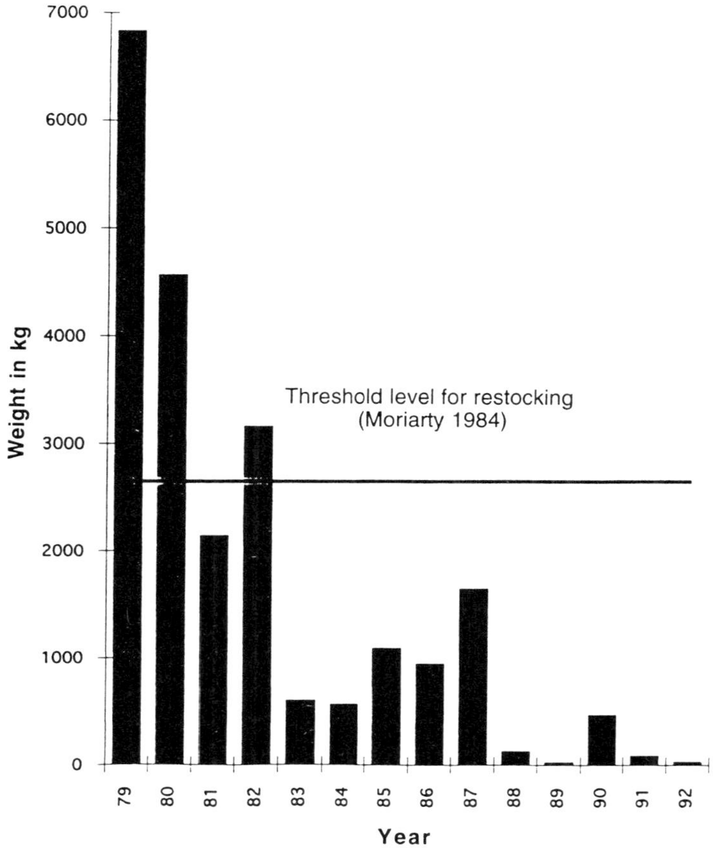

Fig. 1—Yearly catch of elvers at Ardnacrusha from 1979 to 1992.

RESULTS AND CONCLUSIONS

The preliminary surveys, using trawls, fixed nets and shore searches, showed promising concentrations of glass eels in the upper estuary, and indicated several sites where glass eel/elver fishing might be conducted. Experimental fishing for glass eels was then performed at night on the incoming tide, using two methods.

(i) The Botirão net, a Portuguese glass eel net which consists of a funnel, with two large wings attached, is anchored to both sides of the river with the funnel facing into the incoming tide so that it is pushed open by the current. The incoming tide is fished and the funnel is then closed and emptied of its catch.

(ii) Hand nets and conical pull nets are swept into the water against the rising and falling tide.

Numbers of glass eels have been caught in and near tributary mouths using both these methods, with concentrations markedly affected by stage of tide and time of year. This work is continuing, but preliminary results indicate that the Shannon has the potential to support a small commercial style glass eel fishery for the purpose of restocking the lakes. Alternatively, some of the glass eels caught could be kept for a period of ongrowing and the resultant larger elvers could then be used for restocking the lakes. This would help to decrease juvenile mortality rates.

The numbers of glass eels used for stocking would only be a small proportion of the numbers entering the system and therefore would not have a detrimental effect on organisms which rely on the glass eel as a food source. It is hoped that, over time, the stocked juvenile eels will increase the numbers of resident brown eels and silver eels migrating, probably also increasing the number of escapements at the silver eel fishing sites. Such action is important, as eel populations are declining throughout Europe.

ACKNOWLEDGEMENTS

We should like to thank the Electricity Supply Board, Hydro Group, for its sponsorship of this research, and Dr J.D. Reynolds, project director, for his assistance in preparation of the manuscript.

REFERENCE

MORIARTY, C. 1984 Stock enhancement in the Irish eel (*Anguilla anguilla)* fishery. *EIFAC Technical Papers 42 (*suppl. vol. 1), 50–5.

In: J.D. Reynolds (ed.) 1996 *The conservation of aquatic systems*, 167–170. Dublin. Royal Irish Academy.

THE PRESENT STATUS OF THE OYSTER FISHERY OF LOUGH FOYLE

F.B. McKelvey and T.E. Andrew

Department of Biological and Biomedical Sciences, University of Ulster, Coleraine

ABSTRACT

Lough Foyle, an estuary between counties Donegal and Derry, supports an important stock of native oysters *Ostraea edulis* of recently recognised economic importance. The fishery is one of the few native fisheries in Western Europe still uncontaminated with the parasitic disease *Bonamia.*

The lough, an extensive shallow estuary is 0.5–7.5m deep in most areas. Natural oyster beds are situated mainly in the centre of the lough and along the western (Donegal) shore. Natural recruitment to the fishery is irregular and has been enhanced by artificial stocking. The fishery is potentially at risk from overfishing, pollution and introduced disease. Individual and population growth and recruitment are controlled by environmental temperature, food availability, spat settlement and predation. A research programme will monitor the oyster populations and the ecological factors influencing population conservation.

INTRODUCTION

Lough Foyle, which lies between counties Donegal and Derry, is an extensive shallow estuary covering an area of approximately 200 km^2 (Fig. 1). The lough supports significant stocks of salmon, whitefish and shellfish. Its importance as a shellfishery has only recently been recognised and consequently the Foyle Shellfish Co-operative Society has been formed to promote its development. The native oyster fishery, although important, is only part of the shellfishery. Other species fished include mussels and, to a lesser extent, lobsters, crabs, winkles, cockles. As a whole the fishery supports between 80 and 100 boats, with on average about 300 people fishing plus shore support. The location of the primary oyster beds is shown in Fig. 1. Although these are the major beds, other smaller areas around the lough have been identified.

Most of the oysters are sold through local dealers. The principal markets

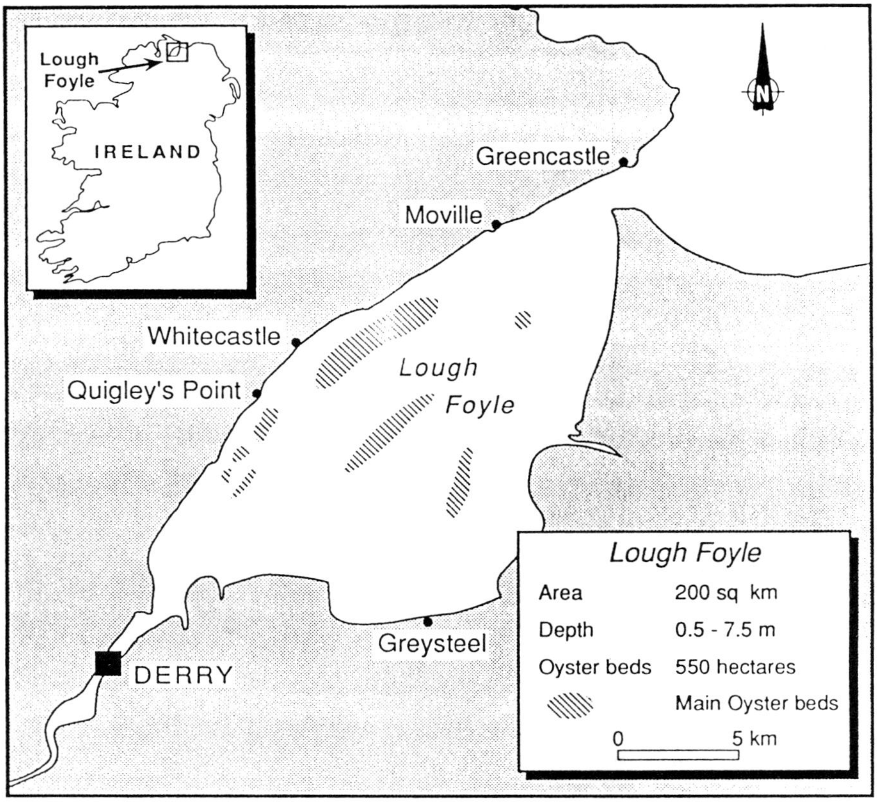

Fig. 1—Map showing the location and extent of oyster beds in Lough Foyle.

are in Britain and France, whose stocks have been greatly reduced through disease, although oysters are also exported to destinations such as Hong Kong.

LIFE CYCLE OF THE NATIVE OYSTER

The principal biological events in the life cycle of *Ostrea edulis* are attainment of sexual maturity of both male and female, successful spawning, larval release and spat settlement.

In natural waters spawning of *O. edulis* does not usually begin until the water temperature reaches about 16°C (O'Connor *et al.* 1992). Fertilisation occurs within the oyster and is followed by development to fully shelled larvae. After a brooding period of one or two weeks larvae are released and they swim through the water column in search of suitable settlement substrates. The larvae settle to become spat, and metamorphosis continues until the adult structure has been developed (Walne 1974).

This project is concerned with the biology and ecology of native oysters *O. edulis* rather than with commercial aspects. Although currently in its preliminary stages, it has particular relevance to the following areas.

Population density and structure

It is relatively difficult to get a measure of the density of the population in a commercial environment; however, estimates are possible. The efficiency of the sampling device is important when making such estimates. Edwards (1980) showed that mean dredge efficiency decreased with reduction in oyster size and for small individuals was approximately 5%. Good estimates of age structure are possible for individual oyster beds using representative samples. The histogram in Fig. 2 indicates the current population size structure of *O. edulis* in Lough Foyle.

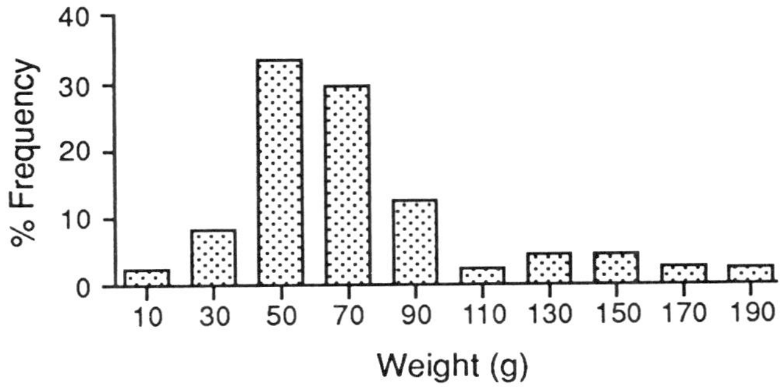

Fig. 2—Histogram indicating the current size structure of the oyster populations of Lough Foyle.

Recruitment

Lough Foyle shows a good recruitment only every 4–5 years (McKelvey *et al.* in prep.). It is thought that this may be due to the low temperature of the lough, which only occasionally rises to about 16°C (temperature suitable for spawning) and falls again very quickly. Once spawning has occurred, other factors will affect recruitment.

Fecundity and survival of larvae. Older oysters produce a higher number of larvae with a higher survival rate than do young oysters (Cunningham 1991).

Settlement surfaces. Young oysters settle on old shell (cultch), tiles or plates; however, the degree of fouling of the settlement surfaces is important. An essential aspect in the provision of settlement surfaces in Lough Foyle is the fishing practice of the fishermen. After the dredge has been towed for a short time, the samples are taken on board and immediately sorted. Undersized oysters and cultch are returned and in this way bed area is increased. During dredging old shell is pushed ahead creating even bigger settlement areas.

Predation. When considering recruitment, predation of the oysters is important. In Lough Foyle the major predators are crabs (shore and hermit) and starfish.

Competition. Potentially more important than predation is the competition from significant mussel populations in the beds. Mussels have a number of effects on oysters in shared ground through competition for food and space. They also cause fouling by the creation and expulsion of large quantities of waste matter.

Disease

The major cause of native oyster disease and mortality in western Europe is the parasite *Bonamia.* Britain, France and Holland have been worst hit. *Bonamia* affects the individual oyster by gradually weakening it until it dies. It has also been noted that it quickly leads to increased susceptibility to predators. In southern Ireland stocks have been reduced by the effect of the parasite. To date, Lough Foyle stocks have not been infected by this disease and it is imperative for the survival of the fishery that it is not introduced by, for example, the importation of infected shellfish.

Pollution

Pollution in Lough Foyle does not pose a major threat to the shellfishery. However, any spillages that may occur must be treated with caution as shellfish, because of their filter feeding mechanism, have the ability to concentrate substances in their body at much higher levels than the surrounding water.

A research programme is currently under way to investigate the above factors in relation to Lough Foyle. The aims of the project are to survey the natural oyster populations and their environment and to monitor any changes; to examine recruitment in the lough and determine how optimisation may be achieved; and to provide information regarding the stocks which may be used in establishing and developing a management structure for the resources.

REFERENCES

CUNNINGHAM, P. 1991 The Lough Foyle oyster fishery, 1990. Commissioned unpublished report. Dublin. B.I.M.

EDWARDS, E. 1980 *An evaluation of the Tralee Bay oyster fishery.* Resource Record Paper. Dublin. B.I.M.

McKELVEY, F.B., ANDREW, T.E. and CUNNINGHAM, P. 1993 Recruitment potential of native oysters *Ostrea edulis* in Lough Foyle. Report to ICES 81st Statutory Meeting, September 1993.

O'CONNOR, R., WHELAN, B.J. and CRUTCHFIELD, J.A. (with O'SULLIVAN, A.J.) 1992 *Review of the Irish aquaculture sector and recommendations for its* development. Dublin. Economic and Social Research Institute.

WALNE, P.R. 1974 *Culture of bivalve molluscs: 50 years' experience at Conwy.* West Byfleet Fishing News (for the Buckland Foundation).

In: J.D. Reynolds (ed.) 1996 *The conservation of aquatic systems*, 171–180. Dublin. Royal Irish Academy.

IMPACT OF ACIDIFICATION ON THE ECOLOGY OF UPLAND STREAMS, WITH PARTICULAR REFERENCE TO THE POSSIBLE EFFECTS OF PLANTATION FORESTRY

M. Kelly-Quinn, D. Tierney and J.J. Bracken

Department of Zoology, University College, Dublin

ABSTRACT

In 1990 a national investigation into the possible effects of plantation forestry on the hydrochemistry, hydrology and ecology of salmonid streams was initiated. The project (AQUAFOR) is an ongoing interdisciplinary effort involving three universities, each working in different parts of the country: County Wicklow (University College Dublin), Galway/Mayo (Trinity College Dublin) and the south-western region (University College Cork). This paper reports on the work which was carried out in County Wicklow. Intensive monitoring of the hydrochemistry of the poorly buffered headwater streams of the Liffey, Kings and Avonmore rivers, as well as the feeder streams of the Vartry Reservoir, was undertaken. Macroinvertebrate and trout populations were also quantified. Sources of acidity, which may include organic acids, non-marine sulphate and nitrates, have been identified and where possible related to the ecology of these systems.

INTRODUCTION

Extensive research has been carried out on the acidification of surface waters in Europe and North America. There is considerable evidence that recent changes in the pH of many rivers and lakes have been caused by deposition of atmospheric pollutants containing compounds of sulphur and nitrogen (Seip and Tolan 1978). It is well known that the susceptibility of freshwaters to acidification is determined by a large number of factors. These include the quality and quantity of atmospheric deposition, the acid neutralising capacity and weatherability of the solid geology, the base status and depth of the overlying soils, and land uses in the catchment.

Evergreen or plantation forestry has been shown to exacerbate the acidification processes in soft water streams draining areas which receive heavy loads of atmospheric pollutants. The generation of acidity has been linked to the ability of the coniferous canopy to scavenge and concentrate airborne pollutants (Harriman and Morrison 1982). Although Ireland has relatively low levels of atmospheric pollution, recent studies report that some freshwater systems on the east coast receive moderate levels of artificial acidification (Bowman 1991). Allott *et al.* (1990) reported an increase in acidity associated with the afforestation of poorly buffered streams in Connemara and south Mayo. These findings, coupled with the fact that Irish forestry is rapidly expanding, and that afforestation was largely concentrated on the poorer upland soils in the catchments of important salmonid nursery streams, created a great deal of concern among fisheries interests. This concern was the impetus for the initiation of a national investigation into the possible effects of plantation forestry on salmonid waters.

The work, which commenced in 1990, is an interdisciplinary, cooperative effort involving three universities working in different parts of the country: County Wicklow (University College Dublin), Galway/Mayo (Trinity College Dublin, project leader N. Allott, Environmental Science Unit) and the south-western region (University College Cork, project leader Paul Giller, Zoology Department). The results of a two-year study are currently being compiled by each group. Papers dealing with the west of Ireland have been published (Allott and Brennan, in press; Reynolds *et al.* in press).

This paper presents an overview of the findings of the UCD group which was working in County Wicklow, one of the most heavily afforested counties in the country.

SITES, MATERIALS AND METHODS

Sites were located on the headwater tributaries of the Kings, Liffey and Avonmore rivers, and on the feeder streams of the Vartry Reservoir (Fig.1). During the two-year study period 63 sites on 33 rivers were investigated. These waters are typical trout nursery streams, characterised by short riffle-glide-pool sequences. Torrential conditions are typical of the majority of the sites, with water levels rising by up to 1m during wet weather. Catchment characteristics were compiled from maps and landsat images using a Geographical Information System. With the exception of two sites on the lower reaches of the Kings and Liffey rivers, all catchments were less than $20km^2$. Average stream width varied from 0.5m to 12m, but the majority of sites were less than 4m wide. Vegetation cover ranged from open moorland to closed canopy plantation forestry.

The solid geology of the area is relatively simple, consisting of a large central granite batholith surrounded by Palaeozoic sedimentary rocks, such as shales. The contact between the two rock groups is marked by a metamorphic aureole consisting principally of mica schist. Soils are varied, but in general the granite has overlying peat or peaty podzols, whereas acid brown earths and podzolics are more typical of the remaining areas. Glacial activity has resulted in moderately deep mineral soil in the eastern part of the county.

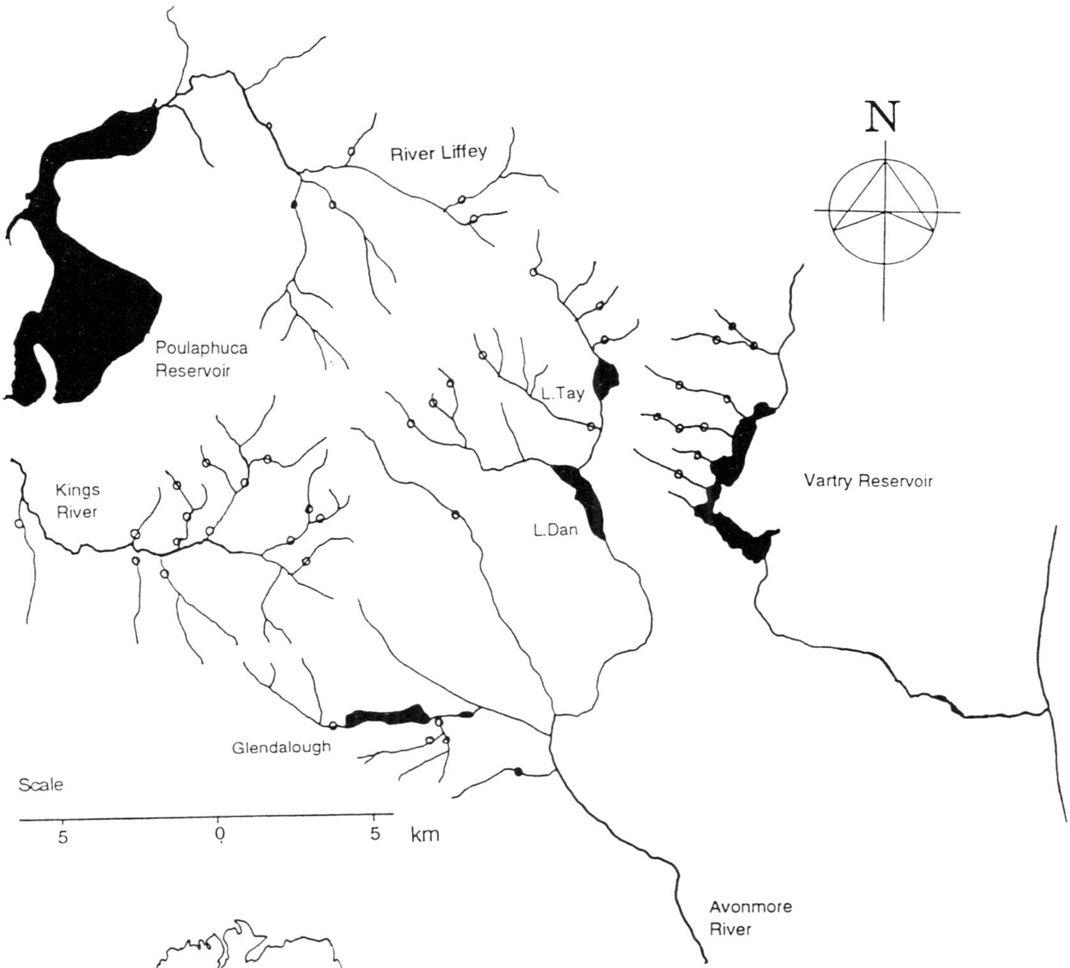

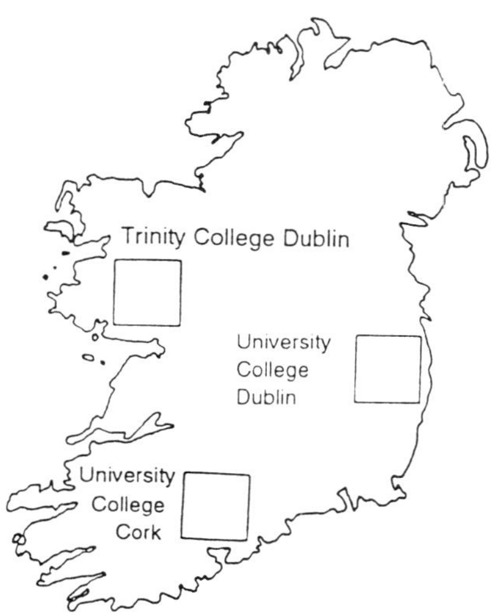

Project centres and universities involved in the national study on the possible impact of afforestation on salmonid streams

Fig. 1—Location of sampling sites in County Wicklow.

Standard hydrochemical and biological methodologies were adopted by the three universities. These have been described by Allott and Brennan (1993) and Reynolds *et al.* (1994). Hydrochemical sampling began in June 1990. Twenty-five sites were sampled intensively at weekly to monthly intervals from October to April and less frequently during the rest of each year. In order to identify the cause of acidity in specific 'acid' episodes, automatic Sigma 800SL water samplers, programmed to collect samples every one to two hours, were installed on the Annalecka (33.3% mature coniferous forest cover) and Ballinagee (moorland) streams in March 1991. These two sites were on adjacent catchments with similar geology and general morphometry and were the most suitable sites available for this comparative work. The remaining 30 sites were sampled at least once every season. Sampling continued until May 1992.

Electrofishing was carried out at 36 locations between July 1990 and July 1992. Standard population density (two or three catch depletion) and growth estimation methods were employed.

An extensive collection of riffle samples was taken at 50 sites (5 replicate Surber samples per site) in April/May 1990. With the exception of Chironomidae and Oligochaeta, invertebrates were identified to the lowest possible taxon and densities were also recorded.

RESULTS

The systems examined are low-ionic strength waters with conductivities not exceeding 90μS on granite and 120μS on Palaeozoic sediments. Table 1 compares the hydrochemical characteristics of moorland and afforested streams on granite.

Table 1—Hydrochemical characteristics of moorland and afforested streams on granite bedrock.

Site	*pH*	*H^+*	*Conductivity*	*Alkalinity*	*LM Al*	*Cl^-*	*$NMSO_4^{2-}$*
M	4.26–7.52	0.03–55	32–56	–70 –+262	0–49	51–256	9–78
A	3.7–7.51	0.03–200	39–80	–154 –+218	0–587	53–411	13–129

(Units: conductivity μS, labile monomeric aluminium (LM Al) μgl^{-1}, non-marine sulphate ($NMSO_4^{2-}$) $\mu\ eql^{-1}$)
M = moorland streams; A = afforested streams

Most streams had pH values above 6.0 during low water conditions. With the exception of two streams in the Vartry catchment, where pH did not fall below 6.0 during the entire study period, all other streams examined were episodically acidic with high H^+ ion values. On granite bedrock H^+ ions were correlated with increasing forest cover (r^2=0.828). Localised variations in the geology, such as the presence of calcite in the granite, appear to increase the buffering capacity of a number of streams. Associated with increasing stream acidity was an increase in the concentrations of aluminium. The proportion of labile monomeric aluminium increased with increasing forest cover. Figure 2 compares the levels of total and labile monomeric aluminium in a moorland and a heavily afforested site in the Kings catchment. A comparison of afforested and moorland streams exhibiting similar pH values revealed that a higher proportion of the total aluminium

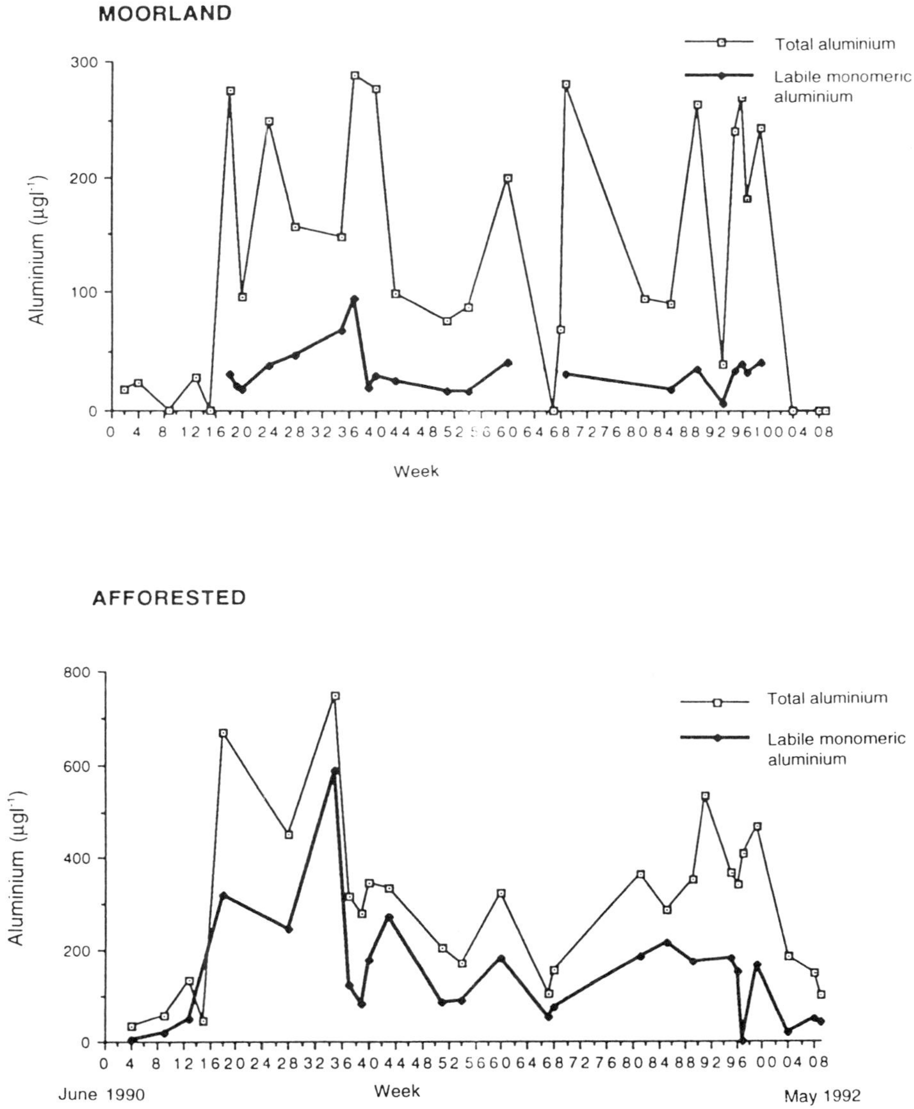

Fig. 2—Total and labile monomeric aluminium concentrations recorded in a moorland and afforested stream between June 1990 and May 1992.

was in the labile form in afforested streams. On mineral soils (afforested) this value was even higher, with labile monomeric aluminium accounting for 75–90% of the total aluminium present.

The concentrations of the conservative chloride anions and of non-marine SO_4^{2-} were also higher in the afforested systems. Buffering capacity in these upland streams draining granite bedrock was very low, as evidenced by the low alkalinity values, which were negative on many occasions. Again, heavily afforested catchments tended to exhibit lower alkalinities than their equivalent moorland streams. Poor buffering capacity is a reflection of the low ionic composition of these waters. Mean calcium and magnesium concentrations were generally below 150m eql^{-1} and 75m eql^{-1}, respectively. Most of the calcium was catchment-derived, but over 50% of the magnesium was of marine origin.

The 'acid events' on the afforested streams tended to give lower pH/higher H^+ ion values and be of longer duration than the control moorland site. Table 2 presents details of four typical acid episodes on the moorland and afforested sites. Organic acids, non-marine SO_4^{2-} and nitrates appear to be the most important contributors to the acidity reported. There is a high natural background acidity which appears to be largely caused by organic acids. While increases in organic acids appear to cause a small increase in H^+ ion, they were nevertheless a major contributor to the total acid concentration in the majority of the samples. However, small increases in non-marine SO_4^{2-} resulted in substantial increases in H^+. This was most severe in the afforested stream, presumably as a result of scavenging of sulphate by the coniferous forest canopy. High non-marine SO_4^{2-} inputs were the result of easterly air movements. Allott and Brennan (1993) also reported that non-marine sulphates were the major source of acidity in afforested catchments in Galway/Mayo. However, in contrast to the west of Ireland, increases in nitrates, whether derived from atmospheric deposition or soil processes, may also have had an influence on the higher acidity of the afforested stream in County Wicklow. High nitrate values were associated with high sulphates on a number of occasions, e.g. on 3 March 1991. Hydrological events also appear to have influenced the changes in pH/H^+. If non-marine hardness is taken as an indication of buffering capacity, then a decrease in hardness resulted in poorer buffering and associated increases in acidity.

During many of the acid episodes the afforested stream recorded large decreases in catchment derived hardness (e.g. 25 March 1992). One hypothesis is that rapid runoff, via the drainage network of the afforested catchment, leads to reduced contact with soils for buffering mechanisms. However, other factors such as catchment size and slope characteristics must be taken into consideration. The Annalecka catchment had a smaller area than that of the Ballinagee stream. Analysis to date tends to indicate that catchment size is unimportant in determining stream acidity but that percentage forest cover is the major controlling factor. Similar conclusions were reported by Allott *et al.* (1990). Further support for forest mediated acidification is obtained when the pH of samples taken along the course of

Table 2—Hydrochemical details of 'acid episodes' in an afforested (A) and moorland (M) stream. On each sampling date the upper set of figures represents hydrochemical conditions prior to the acid episode. The lower set of data are conditions at the peak of the event. With the exception of total organic carbon(TOC), which is expressed in mgl^{-1} and pH, all other values are in μ eq l^{-1}.

Site	*Date*	*Time*	H^+	*pH*	*NM* SO_4^{2-}	NO_3^-	Cl^-	*NM* Na^+	*NM Hardness*	*TOC*	A^-
A	3.3.91	13.17	68	4.17	39	40	199	43	89	10	58
	3.3.91	20.17	129	3.89	64	50	206	18	56	11	55
M	3.3.91	13.30	16	4.79	22	10	143	47	72	96	2
	3.4.91	24.30	25	4.61	42	16	175	41	67	96	4
A	17.12.91	10.30	14	4.84	60	17	252	–15	65	12	88
		22.30	76	4.12	81	19	256	–57	38	15	82
		23.30	79	4.10	85	17	235	–9	60	15	87
M	17.12.91	10.24	3	5.50	56	13	241	–32	74	12	100
	18.12.91	14.24	14	4.89	53	13	198	–6	81	14	102
A	25.1.92	11.31	2	5.66	43	22	202	2	55	7	62
		19.31	39	4.41	60	20	210	–7	62	13	82
M	25.1.92	10.37	1	5.99	39	13	198	17	72	6	55
		17.37	5	5.31	47	17	220	–8	82	11	90
A	12.2.92	9.00	11	4.95	44	12	269	–51	60	10	74
	12.2.92	15.00	83	4.08	66	16	208	6	52	14	77
M	12.2.92	10.00	2	5.72	26	9	190	3	83	10	89
	12.2.92	16.00	21	4.68	46	11	198	–12	78	12	84
A	25.3.92	12.15	22	4.65	38	12	232	–39	43	11	76
	26.3.92	6.15	60	4.22	31	8	182	–29	19	15	91
M	25.3.92	11.40	2	5.67	23	7	117	–49	35	6	55
		17.37	5	5.31	47	17	220	–8	82	11	90
	26.3.92	1.40	16	4.80	25	8	175	–23	45	13	92

the stream are examined. Generally, pH values in these upland streams rise with increasing distance from the source. In contrast, a site located 1km upstream of the automatic water sampling site on the Annalecka river was less acidic on most sampling occasions. A similar result was reported on the Garyknock stream (Kings River catchment). The catchment of the top site was one third the size of the other site. As regards slope, the moorland catchment is much steeper than the afforested stream, which suggests that runoff would be faster from the moorland site.

FISH POPULATIONS

The only salmonid encountered was *Salmo trutta* L. Minnow, stoneloach and sticklebacks were recorded at some sites in the Vartry catchment. The trout examined are slow growing fish, reaching a length of 4–8cm at the end of year one and 8–15cm at the end of year two. With the exception of the Shankill river and a number of sites in the Vartry catchment, trout populations were well below 0.5 fish/m^2, which is to be expected in streams draining acid geology. Above 270m populations fell below 0.1 fish/m^2. Trout were recorded as high as 460m, but they were not generally found in stretches of stream higher than approximately 400m. Initial observations indicate that, in addition to altitude, stream gradient is an important factor influencing longitudinal distribution of trout in an upland stream. Six non-afforested sites, all above 400m, had no trout at the time of sampling. Of the afforested sites only nine (five streams) of the nineteen sites fished were fishless. Three of these streams would be expected to carry trout. A small number of the afforested streams carried fish populations below the densities expected of such waters. There has been little change in fish densities over the sampling period with the exception of one stream, near Garyknock, at the Wicklow Gap. Here populations fell from an average of 0.2 fish/m^2, a value typical of upland streams, to below 0.05 fish/m^2 in 1992.

Eighty-two taxa were recorded but no site had more than 50% of this figure. Only three sites supported gammarids. They were in the Vartry catchment, where pH values were generally above 6.0. The average number of invertebrates increased with distance from the source. This was probably a reflection of improving pH and less torrential conditions. Ten ephemeropteran species have been identified. *Baetis rhodani* was the most common mayfly recorded. Two species—*Ameletus inopinatus* and *Siphlonurus lacustris*—have been found in small numbers in the high altitude sites. Plecoptera were more abundant than Ephemeroptera at the majority of sites. Seven species were recorded, the most common being *Leuctra inermis*, *Protonemura meyeri* and *Amphinemura sulcicollis.* Table 3 compares the average number of taxa recorded in moorland and afforested streams at two different altitude ranges.

Table 3—Numbers of macroinvertebrate taxa recorded in moorland and afforested streams at different altitude ranges.

Stream type	*Vegetation*	*pH range*	*Labile monomeric Al*	*Number of Taxa*
High altitude				
>350m	Moorland	4.26–7.5	0–49	14(9–22)
	Afforested	4.0–7.5	0–150	10(7–17)
Lower altitude				
<350m	Moorland	4.3–7.52	0–47	30(19–36)
	Afforested	3.7–7.51	0–587	18(8–33)

Labile monomeric aluminium in mgl^{-1}

At the higher altitude there was little difference in the diversity of biota recorded at moorland and afforested sites. Sites were not heavily afforested at this altitude. In the lower reaches of the streams afforested sites had a poorer macroinvertebrate fauna than comparable moorland sites. The lowest numbers of taxa were recorded in two heavily shaded afforested sites. Streams with the highest forest cover had an impoverished fauna dominated by Plecoptera. These sites did not support Ephemeroptera and were also listed among the fishless sites.

DISCUSSION

The poorly buffered streams described in this paper are delicately balanced in their ability to neutralise acidity. They also support aquatic biota that are living towards the lower limit of their pH tolerances. The low diversity of species reported is typical of acid waters and the same pattern has been reported in other similar studies (Harriman and Morrison 1982; Sutcliffe and Hildrew 1989; Reynolds *et al.* 1992). When increases in acidity are not readily neutralised by calcium and magnesium, H^+ and aluminium ions make up the balance. The combination of high H^+ ion, high levels of labile monomeric aluminium and low calcium is not tolerated by salmonids and many macroinvertebrates such as the Ephemeroptera. The resulting impoverished fauna may be caused by physiological stress induced by these chemical conditions. Alternatively the chemical conditions may affect invertebrates indirectly via their food chain (Sjoström 1990). It is often difficult to isolate the influence of such factors and both may interact to determine the composition of the aquatic community (Sutcliffe and Hildrew 1989). In episodically acidic streams, organisms must be able to tolerate changing water chemistry. Afforested streams often experience sudden changes in pH with water remaining acidified for longer periods than moorland streams. The literature suggests that freshwater organisms can tolerate small changes in hydrochemistry for short periods but that the rapid and prolonged changes associated with some afforested systems result in an impoverished fauna (Reader and Dempsey 1989).

In summary, the findings of this study highlight the susceptibility of some soft water streams to acidification. They also further support the suggestion of Allott *et al.* (1992) that, in the interest of the conservation of these waters for their fishery value, macroinvertebrate diversity or other water use, a policy of non-afforestation of these systems should be adopted.

ACKNOWLEDGEMENTS

The project was supported by funding from Eolas, Coillte Teoranta (The Irish Forestry Board) and a Unilever Newman Scholarship.

REFERENCES

ALLOTT, N.A., MILLS, W.R.P., DICK, J.R.W., EACRETT A.M., BRENNAN, M.T., CLANDILLON, S., PHILLIPS, W.E.A., CRITCHLEY, M. and MULLINS, T.E. 1990 *Acidification of surface waters in Connemara and south Mayo—current status and causes.* Dublin. duQuesne Ltd.

ALLOTT, N.A. and BRENNAN, M. 1993 Impact of afforestation on inland waters. In C. Mollan (ed.), *Water for life.* Dublin. Royal Dublin Society

BOWMAN, J.J. 1991 *A limnological study of Lough Maumwee and Glendalough Lake Upper with special emphasis on acidification.* Ph.D. thesis, National University of Ireland.

HARRIMAN, R. and MORRISON, B.R.S. 1982 Ecology of streams draining forested and non-forested catchments in an area of central Scotland subject to acid precipitation. *Hydrobiology* 144, 37–49.

READER, J.P. and DEMPSEY, C.H. 1989 Episodic changes in water quality and their effects on fish. In R. Morris, E.W. Taylor, D.J.A. Brown and J.A. Brown (eds), *Acid toxicity and aquatic animals*, 67–83. Cambridge University Press.

REYNOLDS, J.D., COOKE, D., SIMON, N., ALLOTT, N.A. and BRENNAN, M. 1994 Macroinvertebrate communities of poorly buffered afforested and non-afforested streams in western Ireland. *Verhandlungen Internationale Vereinigung für Theoretische und Angewandte Limnologie* 25, 1496–1501.

SEIP, H.M. and TOLLAN, A. 1978 Acid precipitation and other possible sources for acidification of rivers and lakes. *The Science of the Total Environment* 10, 253–70.

SJOSTRÖM, P. 1990 Food and feeding of *Baetis rhodani* (Ephemeroptera) in acid environments. In B.J. Mason (ed.), *The surface waters acidification programme,* 427–30. Cambridge University Press.

SUTCLIFFE, D.W. and HILDREW, A.G. 1989 Invertebrate communities in acid streams. In R. Morris, E.W. Taylor, D.J.A. Brown and J.A. Brown (eds), *Acid toxicity and aquatic animals,* 13–29. Cambridge University Press.

In: J.D. Reynolds (ed.) 1996 *The conservation of aquatic systems*, 181–183. Dublin. Royal Irish Academy.

CLIMATE CHANGE AND THE IRISH AQUATIC FAUNA

J.N.R. Grainger

Department of Zoology, Trinity College, Dublin

ABSTRACT

A brief summary is given of the likely effects of a climatic temperature rise on the fauna of various aquatic habitats. It is suggested that southern forms may extend their distribution northwards and northern species may retreat. Cold stenotherms may have to move further upstream in streams. The effects may be small on some inhabitants of temporary waters. The rising sea level may have a major effect on estuarine habitats, mud flats and salt marshes, as well as on the intertidal fauna of open coasts. It is stressed that the lack of distribution information on many species is very limiting in deciding what the effects may be on aquatic faunas.

INTRODUCTION

Assessing the likely effects of a climatic temperature rise on the aquatic fauna of Ireland is difficult because of the uncertainties involved. These arise from two main sources. Firstly, while climatic warming is certainly taking place on a global scale, the likely scale of change in Ireland is a matter of speculation. Secondly, the lack of distribution information on species in many groups in Ireland is a very serious limitation.

The life cycles of most invertebrates is short (frequently less than a year) and so they are greatly influenced by weather. Populations, on the other hand, result from successive generations and these show the effects of climate. Thus species at the edge of their distribution in Ireland might be expected to show the greatest changes. This is why the lack of distribution information for many groups is so serious.

Climatic warming means that the chances of low winter temperatures occurring are reduced, and similarly the chances of higher summer temperatures are increased. This was discussed in Grainger (1992). There have been several recent publications on climate change in Ireland, including possible effects on the Irish fauna (Grainger 1992; McWilliams 1991). The papers in McWilliams (1991) are based on the assumptions made by the

Intergovernmental Panel on Climate Change (IPCC) Working Group I (Houghton *et al.*1990). This paper will not repeat what has already been discussed in the papers mentioned above but will deal briefly with the effects on the main aquatic habitats.

LAKES AND PONDS

Most lakes in Ireland are relatively shallow and few develop stratification in summer. Temporary stratification may develop in some and will be broken down by wind after a time. Since Ireland is a rather cloudy country, especially in the west, water temperature tends, broadly, to follow air temperature; higher air temperatures will therefore result in higher water temperatures. It is difficult to say whether this will have much effect on lake zooplankton and fish populations. George (1991) concluded that the effect on the former might be small. Fish populations may, however, show marked changes, but it may be difficult to disentangle these from the effects of eutrophication, acidification, etc.

STREAMS

A rise in temperature will affect all stream-dweller populations, but to what extent is very obscure. Trout and salmon are not far from their southern limits in Ireland and these natural populations may be affected. Temperature falls with increasing altitude (normally about 0.6°C per 100m) and so some species may have to move upstream if they can. Elliott (1991) gives data on the temperature tolerance of stonefly eggs. Some species are stenotherms and would be expected to move further upstream if temperatures continue to rise. It is thus important to list the present fauna of high mountain streams in Ireland.

TEMPORARY WATER BODIES

Temporary water bodies are very numerous in the west of Ireland. The fauna has been little investigated, but studies so far show some interesting organisms (Duigan and Frey 1987; Grainger 1991; Ali *et al.* 1987). A computer model has been made of the survival of the Anostracan *Tanymastix stagnalis* (Grainger 1991). A re-examination of this suggests that a climatic temperature rise is likely to have little effect on survival. A temperature rise will increase evaporation from such habitats, but this is really a significant factor only between May and August. Most of the floodings take place between July and December and a temperature rise then would hasten the completion of the life cycle, provided food is not limiting.

Northern species such as *Eurycercus glacialis* may retreat northwards or become extinct.

MARINE HABITATS

Southern species such as the sea urchin *Paracentrotis lividus* and the Devonshire cup coral *Ceratophyllum* may be expected to edge further northwards. Northern species will probably retreat.

What will happen to sea level is an important problem. It is rising worldwide by about 1mm per year and is expected to continue to rise (McWilliams 1991). However, this may be affected by increased precipitation on the polar ice caps (Jacobs 1993). The situation is more complicated in Ireland, where the northern part of the island is rising in relation to the south (Carter 1991).

A rising sea level may have a large effect, drowning estuaries, mud flats and salt marshes, with the loss of some important habitats and the consequent major disruption of the fauna, including the intertidal fauna of open coasts.

REFERENCES

ALI, T.H., HOLMES, J.M.C. and GRAINGER, J.N.R. 1987 *Diaptomus cyaneus* Gurney, a freshwater copepod new to Britain and Ireland. *Irish Naturalists' Journal* 22, 240–1.

CARTER, R.W.G. 1991 Sea-level changes. In B.E. McWilliams (ed.), *Climate change —studies on the implications for Ireland*, 125–71. Dublin. Stationery Office.

DUIGAN, C. and FREY, D.G. 1987 *Eurycercus glacialis,* chydorid cladoceran new to Ireland. *Irish Naturalists' Journal* 22, 180–3.

ELLIOTT, J.M. 1991 Aquatic insects as target organisms for the study of the effects of projected climate change in the British Isles. *Freshwater Forum* 1, 195–203.

HOUGHTON, J.T., JENKINS, G.J. and EPHRAUMS, J.J. 1990 *Climatic change—the IPCC assessment.* Cambridge University Press.

GEORGE, D.G. 1991 The influence of global warming on freshwater plankton communities in Britain. *Freshwater Forum* 1, 204–14.

GRAINGER, J.N.R. 1991 The biology of *Tanymastix stagnalis* (L.) and its survival in large and small water bodies in Ireland. *Hydrobiologia* 212, 77–82.

GRAINGER, J.N.R. 1992 The probable effects of climate change on invertebrate growth and reproduction with particular reference to Ireland. *Invertebrate Reproduction and Development* 22, 239–44.

JACOBS, S.S. 1993 Is the Antarctic ice sheet growing? *Nature* 360, 29–33.

McWILLIAMS, B.E. (ed.) 1991 *Climate change: studies on the implications for Ireland*. Dublin. Stationery Office.

In: J.D. Reynolds (ed.) 1996 *The conservation of aquatic systems*, 184–189. Dublin. Royal Irish Academy.

CONSERVATION PROBLEMS FOR AQUATIC SYSTEMS — AN NGO VIEWPOINT

David Hickie

An Taisce—The National Trust for Ireland

ABSTRACT

An Taisce is currently involved with or concerned about unsustainable exploitation of salmonid stocks; forestry, intensive agriculture, quarrying/mining and industry and their impact on freshwater systems; and conservation of aquatic systems of special conservation importance, such as turloughs, freshwater lakes and estuaries.

An Taisce is promoting a Shannon catchment authority; supports sustainable use of aquatic resources (e.g. angling and shellfish farming) where appropriate; and has campaigned for environmental payments for farmers, and ecologically-sound forestry and aquaculture policies.

Eutrophication, physical damage to aquatic habitats and a decline in salmonid stocks demonstrate that current conservation policies are not working properly. There is some cause for optimism with expected legal protection of Areas of Scientific Interest, environmental payments for farmers, and improved guidelines for forestry. However, there is less optimism about the future of wild salmon and sea trout stocks, and urgent conservation measures are needed.

INTRODUCTION

An Taisce is a voluntary body founded in 1948 with the objective of conserving Ireland's natural and cultural heritage. It has about 5500 members nationwide, is governed by a democratically elected council and executive, and employs a small number of professionals and administrators to support its voluntary activities. It is widely recognised that Ireland's natural resources are vulnerable to over-exploitation and damage from certain types of economic development. The existence of An Taisce is due to the need for a voice outside government to articulate the concerns of the public on conservation issues so that progress can be made in achieving conservation objectives. In essence, An Taisce seeks to maintain or, if possible, enhance the natural and cultural resources of the country.

An Taisce is involved with many aspects of conservation of the natural environment, of which aquatic systems form an important part. An Taisce pursues its environmental objectives by means of educational activities, raising public awareness—through conferences, seminars and field trips, and communication with the media—liaison with state agencies and developers, lobbying of decision-makers, and participation in the planning process. An Taisce does occasionally commission scientific research for its own use, but is very reliant on information from third-level and scientific institutions (e.g. the Royal Irish Academy), and state bodies (e.g. the Fisheries Boards, Department of Marine, Wildlife Service, Environmental Research Unit, Geological Survey). If An Taisce and other voluntary organisations are to participate in conservation in a meaningful way, the availability of high quality, up-to-date research information is vital.

ACTIVITIES WITH POTENTIAL IMPACTS ON AQUATIC SYSTEMS

Conventional agriculture

Since the 1960s, there has been a gradual intensification of farming enterprises, and this has coincided with decreasing water quality owing, in large part, to silage effluent and animal slurries entering water courses. Teagasc, the state agricultural development authority, has been providing an advisory service to farmers, which, combined with grants available from the Department of Agriculture for improvements to farmyards, has significantly reduced incidents of gross water pollution in recent years. An Taisce's current response is to avoid highlighting in the media farm pollution from conventional farming, or making public complaints, which would be counter-productive, since genuine efforts are being made to educate farmers about pollution problems; most farmers are now highly aware of farm pollution, and significant sums of money are being spent on pollution abatement. An Taisce has recommended a continuation of the grant scheme for pollution control, using EC Structural Funds, and has publicly commended the efforts made to resolve problems to date. Education and advice on farmyard pollution control are now reasonably good. It will take several more years before remaining farms with high pollution risks are attended to, but it has to be acknowledged that considerable progress has been made since 1989, when grants were first introduced and enforcement of legislation improved.

Large scale factory agriculture

Expansion of large-scale intensive farming poses some serious problems for aquatic systems. Pig installations with over 3000 pigs, or over 1000 pigs on gley soils, have been subject to planning permission and environmental impact assessment since 1990. An Taisce has viewed the expansion of such large units in recent years with concern, and has frequently made submissions to planning authorities on pig developments. In such circumstances, An Taisce will liaise with and seek advice from the Fisheries Boards—whose staff usually have an extensive technical knowledge of aquatic resources in the locality—and hydrogeologists. An Taisce is often called on to respond to proposals for pig units by local communities. Many surface-water bodies, some of which have been recognised as Areas of Scientific Interest, have been degraded by eutrophication in areas where there are high concentrations of intensive farming enterprises. Some regions, such as Cavan/Monaghan and the Lough Sheelin catchment, already have a manure surplus and any further expansion of factory farms may become an unacceptable risk.

Drainage

Arterial drainage has declined significantly in recent years, and is no longer regarded as being high on the agenda of An Taisce. However, the Office of Public Works has not yet abandoned plans to drain the Owenmore and Dunkellin catchments, or the Mulcair and Suir. EC funding would not easily be obtained in the unlikely event that cost/benefit analyses proved positive. The cross-border Finn–Lackey drainage scheme was abandoned in 1990. Piecemeal drainage by local authorities for flood relief or local drainage schemes such as that for Rahasane turlough in east County Galway have caused ecological damage in recent years, and this has been highlighted by An Taisce. Small water bodies are constantly being damaged by agricultural and forestry activities. An Taisce has been highlighting this problem through education of farmers and farm advisers. It has also encouraged or made recommendations for re-creation of ponds on farms.

Forestry

Forestry can have some significant impacts on aquatic systems, particularly in upland areas with peaty soils, oligotrophic lakes, and waters with a low alkalinity and poor buffering capacity. The upper reaches of river catchments are easily disrupted by ploughing, drainage, fertilisation, felling activities and road-making. Spawning beds for salmon and trout have been damaged, some water bodies have suffered from algal blooms and a few fish kills have occurred. Some of the impacts on aquatic systems resulting from forestry activities may be temporary in nature, others may have more lasting effects, such as conifer-induced acidification. An Taisce has taken the view that forestry activities do need to be better controlled. Although introducing planning permission for most forestry projects could be counter-productive, projects over 100h in extent should at least be subject to environmental impact assessment (EIA). At present, only projects over 200h are subject to planning permission and EIA.

The recently issued forestry and fisheries guidelines go some way towards providing solutions to conflicting situations and offer a valuable opportunity for consultation between fisheries staff and forestry developers. The guidelines also provide for the designation of sensitive catchments, where extra precautions must be taken to prevent damage to aquatic systems. A drawback at present is that no limits have been placed on the proportion of a catchment that can be afforested without reductions in water quality and damage to aquatic life resulting from conifer-induced acidification. There are certain limited areas, identified by the Fisheries Boards and the Enviromental Research Unit as acid-sensitive (Bowman 1991), where An Taisce believes that no further large-scale planting of conifers should be grant-assisted because of potential risks to water quality from acidification. An experiment in liming of a poorly buffered catchment in the Ox Mountains in County Sligo is planned in conjunction with a Coillte forestry development. It is not yet clear if beneficial results can arise from such exercises and, in general, they should proceed with caution.

Aquaculture

Intensive salmon farming in sea cages has a greater social and environmental impact than shellfish farming, such as rope culture of mussels, intertidal culture of oysters or clams. In general, shellfish are more sensitive indicators of water quality and are more vulnerable to insecticides, drugs and organic waste than salmon in cages. Small amounts of organic waste have been found to increase shellfish growth, but there is some pressure on dischargers of other wastes, such as domestic sewage, to reduce or eliminate their discharges (O'Sullivan 1992).

Scientific evidence that salmon farming increases the numbers of sea lice parasites, which seriously affect wild sea trout (*Salmo trutta*) stocks in the west, is persuasive. Large numbers of juvenile sea lice were found to be attacking sea trout during their delicate transitional phase, from fresh to salt water, creating lesions which exposed the fish to secondary infections or destroyed their ability to osmoregulate. Adverse climatic conditions combined with large numbers of sea lice are now strongly considered to have caused the collapse of stocks in the west region in 1989 and 1990. There was a further bad year in 1991. Salmon production from coastal farms increased dramatically during the latter half of the 1980s, especially in the west region, and this is considered likely to have caused the increase in sea lice numbers far in excess of the natural background levels emanating from wild fish, although there may be other contributory environmental factors. According to the Sea Trout Action Group (STAG), 'the key to restoring and maintaining sea trout stocks is the elimination of the sea lice problem on salmon farms close to sea trout fisheries. Until this is done, the probability of further stock collapses remains high' (STAG 1992). An Taisce's response to the problems posed by certain fin fish farming activities has been to support further research conducted by STAG and STAG recommendations, such as temporary or permanent removal of sea cages in areas affected by excess sea lice.

Drift netting

Drift netting is contributing to the depletion of wild salmonid stocks, and An Taisce has supported recommendations that the state should buy out drift netting operations, thereby appropriately compensating the fishermen involved. There is a good case for this since the price of wild salmon has dropped markedly, and drift net fishermen have found it much harder to make a living. An Taisce has also recommended that research should be conducted into salmon ranching as an alternative to drift netting.

SUGGESTIONS FOR PROGRESS IN CONSERVATION OF AQUATIC SYSTEMS

Amendments to development policies in a number of different sectors can make a significant contribution to conservation measures. The introduction of forestry and fisheries guidelines and the provision of grant aid for farmyard pollution control are examples of beneficial initiatives. Giving legal force to Areas of Scientific Interest and providing grant assistance to landowners in such areas are expected moves that have the support of all voluntary conservation bodies. Some examples of future policy changes that could assist in the conservation of aquatic systems are listed below.

- State buy-outs of drift netting operations—which would appropriately compensate the fishermen involved—in order to conserve wild salmon stocks.
- Temporary or permanent removal of sea cages in coastal areas where sea trout numbers have suffered collapses.
- Introduction of controls on expansion of large-scale pig or poultry units in areas vulnerable to water pollution, possibly by introduction of strict licences.
- Policy to discourage further large-scale forestry in certain water catchments affected by or vulnerable to acidification, which may be exacerbated by forestry operations; environmental impact assessment could be introduced for forestry projects over 100h and also for smaller projects in sensitive water catchments.
- Improved education of farmers in order to conserve existing aquatic systems on farms.

  In-service environmental training of departmental and local authority officials.
- Grants, backed up by good advice, to assist farmers in re-creating water bodies on farms.

- Winding down of all current and planned arterial drainage projects under the auspices of the Office of Public Works, and more control of local authority drainage works.
- Improved screening of development projects eligible for government and EC funding which may impact on aquatic systems (for example, tourism projects).
- Improved funding of research programmes on sustainable use of marine and freshwater resources.
- Recruitment to the Fisheries Boards, National Parks and Wildlife Service, and local authorities of some additional professional staff with biological training.

REFERENCES

BOWMAN, J. 1991 *Acid sensitive waters in Ireland.* Dublin. Environmental Research Unit, Department of the Environment.

STAG (Sea Trout Action Group) Report 1992. *Sea Trout News,* no. 3.

O'SULLIVAN, A.J. 1992 Aquaculture and the environment. In R. O'Connor, B.J. Whelan, J.A. Crutchfield and A.J. O'Sullivan (eds), *Review of the Irish aquaculture sector and recommendations for its development.* Dublin. Economic and Social Research Institute.

In: J.D. Reynolds (ed.) 1996 *The conservation of aquatic systems*, 190-191. Dublin. Royal Irish Academy.

THE CONSERVATION APPROACH OF THE WILDLIFE SERVICE

J.M. Ryan

National Parks and Wildlife Service, Office of Public Works, Dublin

ABSTRACT

The paper reviews the statutory responsibilities of the National Parks and Wildlife Service for aquatic conservation under the Wildlife Act and various international conventions. It then deals briefly with the ways in which these responsibilities are presently discharged via regulating, management, research and educational activities, and outlines possible approaches to aquatic nature conservation in the context of moves at national and international level towards the concept of sustainable development.

The National Parks and Wildlife Service has statutory responsibilities for aquatic conservation in the Republic of Ireland both under the Wildlife Act (1976) and through international commitments. At present the Wildlife Act affords some protection to most mammals (including the otter but not the mink), all birds associated with aquatic habitats, three amphibian species (the newt is neither considered internationally important, like the frog, nor threatened, like the natterjack toad), and two aquatic invertebrates (the freshwater crayfish and pearl mussel) both of which are discussed elsewhere in this volume. Fish are not protected under the Wildlife Act. The Act also provides for the establishment of nature reserves for the protection of plants and animals, refuges for fauna to protect animal species and their habitats, and wildfowl sanctuaries within which shooting is prohibited. Complementary legislation provides for the control of water pollution, planning in the countryside, environmental impact assessment, and the establishment of Environmentally Sensitive Areas (ESAs).

Responsibilities are also discharged under various international conventions. The Berne Convention (1982) obliges contracting states to conserve a number of plants, animals and their habitats, and to declare Biogenetic Reserves. This convention protects the otter, frog, natterjack toad, eight fish species and a number of wetland birds. The Bonn Convention (1983) provides for the conservation of migratory birds. In addition, the Ramsar

Convention (1991) seeks to protect listed wetland habitats of particular importance for waterfowl—21 such sites have been designated in the Republic. Finally, Ireland has responsibility for aquatic mammals under various Whaling Conventions.

Under EC legislation, the Birds Directive (1981) emphasises conservation of habitats through Special Protection Areas (SPAs) (21 in the Republic), and the Habitats Directive (1992) lists a range of species and landscape features for conservation. These responsibilities are presently discharged through regulations, management actions, research, and educational activities. There are 1824 Areas of Scientific Interest (ASIs) but, as these were found in 1991 to have no legal basis, all are now being resurveyed and will be considered for designation as Natural Heritage Areas (NHAs). Landowners must be notified of such designation, and can only appeal on a scientific basis.

Habitat information already exists for 1585 ASIs; of these, 224 (14%) are predominantly aquatic (7 marine, 10 brackish and 211 freshwater), and 887 (56%) are part wetland. It is thus clear that the majority of ASI sites listed have some wetland interest.

Of the 1585 ASIs, some 72 are designated statutory Nature Reserves; 13 are marine reserves, one—the Wexford Wildfowl Reserve—is brackish, and 22 are freshwater (15 bogs, 2 fens, 3 lakes, 1 turlough and 1 river). The 2% of land covered by freshwaters in Ireland and the approximately 18% of bogland are thus represented in the network of National Nature Reserves. However, there is a strong case for more aquatic reserves, in particular more lakes. It is hoped that aquatic nature conservation will be further aided by moves at national and international level towards the concept of sustainable development.

In terms of research and education, marine conservation is being furthered by the BIOMAR project, funded by the EU to the Wildlife Service and involving a database and habitat survey, with the participation of Trinity College Dublin and University College Galway, plus scientists from Newcastle on Tyne (UK) and the Netherlands. The Wildlife Service monitors, through its own scientists or through contracts, the status of a number of animals and plants considered vulnerable, and the information is included in Red Data Books. There are ongoing classification exercises for lakes, turloughs and rivers, but at present the Wildlife Service has no active definition of water quality and relates to definitions of the Department of the Environment (see McGarrigle, this seminar). The Service also monitors planning applications that may have an adverse effect on the environment; in 1992, there were 137 applications for afforestation, and the Service objected to 20 of these which were within ASIs.

In: J.D. Reynolds (ed.) 1996 *The conservation of aquatic systems*, 192–194. Dublin. Royal Irish Academy.

CONSERVATION OF AQUATIC SYSTEMS—THE WAY FORWARD

Julian D. Reynolds

Department of Zoology, Trinity College, Dublin

In this seminar, an extensive series of papers and posters explored several aspects of the problem of how best to conserve aquatic systems. Comprehensive reviews of freshwater habitats and water quality and a detailed survey of sublittoral marine species and habitats were followed by regional studies such as the variability of turloughs, the impacts of acidification on Wicklow streams, and the eutrophication of Lough Derg. Synoptic information provided a basis for developing a conservation strategy at Strangford Lough and for assessment of human impacts in Kinsale Harbour.

Studies on species at risk demonstrated the importance of a knowledge of their biology and present distribution. It was emphasised that there are a number of different categories of rarity in the Irish context, each requiring a separate conservation strategy. In this seminar, the general consensus of authors was that both species and habitats may be at risk through environmental change (e.g. temperature, rainfall), as well as through human interference (habitat alteration, pollution, transport of organisms). In many cases more than one cause was suspected.

Most Irish pearl mussel populations are very long-lived, and hence susceptible to both human and environmental influences over time. However, climatic change has not yet been demonstrated for Ireland, and its possible effects are still uncertain. Human impacts are often important: while otters are still widespread in Ireland and tolerant of human disturbance, their continued survival may depend on the control of pesticides and availability of suitable river corridors. Newts appear widespread but uncommon in Ireland; both suitable ponds and undisturbed terrestrial habitats are required for their continued survival. Stocks of freshwater crayfish are widespread, but highly vulnerable to disease introduced apparently with recreational fishing. The sea trout decline appears well correlated with sea-lice prevalence, which may be associated with salmonid aquaculture. The common eel is in long-term decline in terms of both recruitment and fisheries. The identification of new juvenile eel fisheries, whether used for commerce or restocking, should aid conservation and controlled exploitation of yellow eel stocks in freshwater. Activities of the Lough Foyle oyster fisheries help to spread oyster stocks. More generally, a proper knowledge of biology helps to explain fishing impacts on stocks, while monitoring is essential for the conservation of exploited species.

Conservation also has a necessary political aspect. As opportunities permit, the nature reserve network is being expanded, with Natural Heritage Area designation replacing Areas of Scientific Interest. The most profitable approach of most non-governmental organisations is generally to work with government rather than against it, but also to make use of the best available independent research information.

Truth and objectivity must be a priority in presenting a strong scientific case for conservation. Thus, while there is a sense in which all communities are sensitive to external forces, a concerned conservationist may use the terms 'fragile' or 'sensitive' emotively to describe any system or species under threat of exploitation, direct or indirect. The environs of the Burren, for example, are not necessarily fragile or sensitive; the biota of turloughs is adapted to stresses in a number of ways and so may be relatively resilient except to toxic pollutants. Differing human values also affect conservation efforts: while most birds and mammals are widely seen to be precious, it would be difficult to conserve peatlands on the basis of their rare tyrphobiont midge fauna. Finally, there may be conflicting requirements of species each with a valid claim to conservation, such as otter and sea trout; other such conflicts might involve eels, crayfish and otters, or trout and pearl mussels.

The essential pre-requisite to any management decisions or conservation designations is a detailed knowledge of the species and habitats most at risk. There is a need for good available databases, which should include distribution and life history information, position in the food web, etc. This is turn presupposes sound taxonomy, correct ecological information, and active publication of findings. Thus the rise in popularity of SCUBA, and the parallel development of good photographic keys to marine invertebrates, have transformed our knowledge of the marine sublittoral biota down to 50m. Regular monitoring of freshwater quality based on relative abundance of a range of macroinvertebrates has given us a comprehensive database for Irish streams, while detailed information is available for lakes such as Lough Derg, from which their apparent deterioration may be assessed.

CONSERVATION STRATEGIES

Because the situations of rare or threatened organisms are so diverse, no single conservation approach can be proposed, and it is clear that multiple and innovative approaches are needed. In certain instances the strategy of complete legal protection of an organism may be counter-productive, if its net effect is to make that organism invisible to the general public.

In this seminar habitat conservation was generally agreed to be the most effective way in which to conserve areas of scientific interest and their fauna and flora. Even where macro-environmental influences are involved, habitat conservation can be appropriate. If climate change might affect latitude and altitude ranges of important species, we should perhaps be conserving high mountain habitats now, as refuges for the future. However, as types and degrees of rarity differ, so too habitat conservation may not always be an appropriate strategy; it may even be impracticable, especially if the species is both widespread in Ireland and wide-ranging, using different parts of the habitat for different functions; food, shelter or reproduction. For example, otters and crayfish are widespread in Ireland, and penetrate even into urban areas. Habitat conservation may also not protect certain organisms at risk, as evidenced by the fen violet and certain birds and insects, which need some degree of habitat disturbance or interruption of succession in order to breed successfully.

One strategy to arrest the decline of a species or to reduce its vulnerability might be translocation, perhaps to habitats outside its known range. The ethics of translocation and reinstatement of rare species are currently actively under discussion. It is clear that any range extension of a rare species must involve, not only full consultation and appropriate publicity to avoid confusion, but also the most careful consideration of possible effects on the receiving environment, as well as the survival prospects of the animal or plant itself. In the case of rare fish in Britain, such translocations have been made within the same catchment, and generally to water bodies of no obvious scientific value.

For effective conservation, there must also be innovation in legislation. By-laws and guidelines to industry should generally be supported and evaluated, as these are easier to amend than replace. Environmental impact studies should be required for even small agricultural projects.

While there are complex implications of the requirement for free movement of species within the European Union, it is apparent that good management of field equipment belonging to both officials and fishermen would help prevent the unintended transfer of organisms, including pathogens, between states or catchments.

Impacts of commercial activities such as fisheries, mining, land drainage or the application of biocides can obviously have unintended effects on non-target biota. Thus, fisheries may affect non-commercial species, alter community structure, or damage or change habitat. In some situations, however, exploitation may not conflict with conservation priorities.

Finally, funding of conservation may be most effective by funding research. From posters presented at this seminar, it is clear that even a small level of state support can produce valuable data. For larger-scale 'fire-fighting' action, a fund needs to be established, so that rapid effective remedial action may be taken pending longer-term studies.

I hope that this seminar will have stimulated both thought and activity in the conservation of our aquatic resources. Its fruits will only become clear with time, and a re-examination of the status of conservation in this island is in order. In 1998 the largest meeting of freshwater biologists to be held in Ireland will examine all aspects of pure and applied limnology. This seminar will have provided a number of useful benchmarks for their consideration.